T0229531

Artificial Intelligence Enabled Signal Processing based Models for Neural Information Processing

This book provides details regarding the application of various signal processing and artificial intelligence-based methods for electroencephalography data analysis. It will help readers in understanding the use of electroencephalography signals for different neural information processing and cognitive neuroscience applications. The book:

- Covers topics related to the application of signal processing and machine learning-based techniques for the analysis and classification of electroencephalography signals.
- Presents automated methods for detection of neurological disorders and other applications such as cognitive task recognition and brain–computer interface.
- Highlights the latest machine learning and deep learning methods for neural signal processing.
- Discusses mathematical details for the signal processing and machine learning algorithms applied for electroencephalography data analysis.
- Showcases the detection of dementia from electroencephalography signals using signal processing and machine learning-based techniques.

The book is primarily written for senior undergraduates, graduate students, and researchers in the fields of electrical engineering, electronics and communications engineering, and biomedical engineering.

Artificial Intelligence Enabled Signal Processing based Models for Neural Information Processing

Edited by
Rajesh Kumar Tripathy and Ram Bilas Pachori

CRC Press
Taylor & Francis Group
Boca Raton London New York

CRC Press is an imprint of the
Taylor & Francis Group, an **informa** business

First edition published 2024
by CRC Press
2385 NW Executive Center Drive, Suite 320, Boca Raton FL 33431

and by CRC Press
4 Park Square, Milton Park, Abingdon, Oxon, OX14 4RN

CRC Press is an imprint of Taylor & Francis Group, LLC

© 2024 selection and editorial matter, Rajesh Kumar Tripathy and Ram Bilas Pachori; individual chapters, the contributors

ISBN: 978-1-032-52930-1 (hbk)
ISBN: 978-1-032-76765-9 (pbk)
ISBN: 978-1-003-47997-0 (ebk)

DOI: 10.1201/9781003479970

Typeset in Times LT Std
by Apex CoVantage, LLC

Contents

Chapter 12 Detection of Stress Levels During the Stroop Color-Word Test
Using Multivariate Projection-Based MUSIC Domain EWT of
Multichannel EEG Signal and Machine Learning

*Shaswati Dash, Rajesh Kumar Tripathy, Satrujit Mishra,
and Ram Bilas Pachori*

Preface

This book aims to demonstrate applications of artificial intelligence (AI) and signal processing for analyzing and classifying electroencephalography (EEG) and magnetoencephalography (MEG) signals. The topics of this book include combining machine learning methods with signal processing techniques such as transform-domain analysis, time-frequency analysis, multiscale analysis, and feature extraction for detecting neurological disorders such as epilepsy, Alzheimer's disease, and dementia; classifying normal and alcoholic subjects; and recognizing emotions and cognitive tasks.

Chapter 1 discusses recording EEGs, the presence of artifacts, and filtering methods. EEG signal processing applications such as detecting epilepsy, schizophrenia, sleep apnea, and Parkinson's disease; brain–computer interface based on motor imagery EEG; emotion recognition; and cognitive task recognition are discussed. In Chapter 2 and Chapter 4, reviews of AI-enabled signal processing-based approaches is presented for detecting epileptic seizures using EEG signals. Chapter 4 also includes using empirical wavelet transform-based multiscale analysis of EEG signals followed by extracting features from the modes of EE signals. The gradient-boosting models are explored for detecting epilepsy using multiscale features of EEG signals. In Chapter 3, the authors studied using artificial intelligence to classify normal and alcoholic subjects using the features extracted from EEG signals.

In Chapter 5, Chapter 6, and Chapter 7, the authors explore using EEG signal processing and machine learning-based models to recognize emotions. Chapter 7 includes a graph signal processing domain deep neural network-based method for automatically recognizing valence, arousal, and dominance-based emotion classes. Similarly, in Chapter 8, the authors explore using AI-based techniques and the temporal and spectral features of EEG signals to detect Alzheimer's disease. Dementia, a group of cognitive disorders characterized by declining cognitive functions, poses significant challenges to individuals and society; its early detection and intervention are essential in providing the best possible care for affected individuals, improving their quality of life, and managing the societal burden associated with dementia. In Chapter 9, the integration of MEG signals, which provide valuable insights into brain activity with unparalleled temporal resolution, with regularized Riemannian techniques promises to revolutionize the field of dementia screening. This novel approach offers enhanced accuracy and opens the door to a more personalized and noninvasive assessment method.

In Chapter 9 and Chapter 10, the authors explore using AI and EEG signal processing for the automated detection of dementia; under signal processing-based methods, time, frequency, time–frequency, and other features are investigated. Chapter 11 and Chapter 12 are mainly based on the automated recognition of cognitive tasks using EEG signal processing and machine learning techniques. The cognitive tasks induced during the Strop Color-Word Test are recognized from EEG signals using multiscale analysis-based feature extraction and machine learning techniques in Chapter 12.

Editor Biographies

Rajesh Kumar Tripathy received a B. Tech degree in electronics and tele-communication engineering from the Biju Patnaik University of Technology, Odisha, India, in 2009; an M. Tech degree in biomedical engineering from the National Institute of Technology Rourkela, Rourkela, India, in 2013; and a Ph.D. in machine learning for biomedical signal processing from the Indian Institute of Technology (IIT) Guwahati, Guwahati, India, in 2017. He worked as an assistant professor at the Faculty of Engineering and Technology, Siksha 'O' Anusandhan Deemed to be University from March 2017 to June 2018. Since July 2018, he has worked as an assistant professor in the Department of Electrical and Electronics Engineering, Birla Institute of Technology and Science, Pilani, Hyderabad Campus. His research interests are machine learning, deep learning, biomedical signal processing, sensor data processing, medical image processing, and the Internet of Things for health care. He has published research papers in reputed international journals and conferences. He has served as a reviewer for more than 15 scientific journals and as a technical program committee member in various national and international conferences. He is an associate editor of *IEEE Access* and *Frontier in Physiology*.

Ram Bilas Pachori received a B.E. degree with honors in electronics and communication engineering from Rajiv Gandhi Technological University, Bhopal, India, in 2001, and M. Tech. and Ph.D. degrees in electrical engineering from IIT Kanpur, India, in 2003 and 2008, respectively. Before joining the IIT Indore, India, faculty, he was a postdoctoral fellow at the Charles Delaunay Institute, University of Technology of Troyes, France (2007–2008) and an assistant professor at the Communication Research Center, International Institute of Information Technology, Hyderabad, India (2008–2009). He was an assistant professor (2009–2013) and an associate professor (2013–2017) in the Department of Electrical Engineering, IIT Indore, where he has now been a professor since 2017. He is also associated with the Center for Advanced Electronics, IIT Indore. He was a visiting professor at the Department of Computer Engineering, Modeling, Electronics and Systems Engineering, University of Calabria, Rende, Italy, in July 2023; the Faculty of Information & Communication Technology, University of Malta, Malta, from June 2023 to July 2023; the Neural Dynamics of Visual Cognition Lab, Free University of Berlin, Germany, from July to September 2022; the School of Medicine, Faculty of Health and Medical Sciences, Taylor's University, Malaysia, from 2018 to 2019. Previously, he was a visiting scholar at the Intelligent Systems Research Center, Ulster University, Londonderry, UK, in December 2014. His research interests include signal and image processing, biomedical signal processing, nonstationary signal processing, speech signal processing, brain–computer interface, machine learning, and artificial intelligence and the Internet of Things in health care. He is an associate editor of *Electronics Letters, IEEE Transactions on Neural Systems and Rehabilitation Engineering*, and

Biomedical Signal Processing and Control, and an editor of *IETE Technical Review*. He is a fellow of IETE, IEI, and IET. He has 307 publications: journal articles (189), conference papers (82), books (10), and book chapters (26). He has also eight patents, including one Australian patent (granted) and seven Indian patents (published). His publications have been cited approximately 15,000 times with an h-index of 66 according to Google Scholar.

Contributors

Arti Anuragi
Department of Computer Science & Engineering, National Institute of Technology
Raipur, India

Mohd Faizan Bari
Department of Computer Science & Engineering, National Institute of Technology
Raipur, India

Tharun Kumar Reddy Bollu
Department of Electronics and Communication Engineering, Indian Institute of Technology
Roorkee, India

Kritiprasanna Das
Department of Electrical Engineering, Indian Institute of Technology
Indore, India

Shaswati Dash
Department of Electrical and Electronics Engineering, Birla Institute of Technology and Science-Pilani
Hyderabad, India

Fatima Faraz
Department of Medicine, Rawalpindi Medical University
Punjab, Pakistan

Mahbuba Ferdowsi
Department of Mechatronics and Biomedical Engineering, Lee Kong Chian Faculty of Engineering and Science
Selangor, Malaysia

Choon-Hian Goh
Department of Mechatronics and Biomedical Engineering, Lee Kong Chian Faculty of Engineering and Science
Selangor, Malaysia

Shubhangi Goyal
Department of Electronics and Communication Engineering, Indian Institute of Technology
Roorkee, India

Emrah Hancer
Department of Software Engineering, Bucak Technology Faculty
Mehmet, Turkey

Ban-Hoe Kwan
Department of Mechatronics and Biomedical Engineering, Lee Kong Chian Faculty of Engineering and Science
Selangor, Malaysia

Brian Lee
School of Science and Technology, Hong Kong Metropolitan University
Hong Kong, China

Haipeng Liu
Research Centre for Intelligent Healthcare, Coventry University
United Kingdom

Uday Maji
Haldia Institute of Technology
Kolkata, India

Rohan Mandal
Haldia Institute of Technology
Kolkata, India

Hemant Kumar Meena
Department of Electrical Engineering,
    Malaviya National Institute of
    Technology
Jaipur, India

Satrujit Mishra
Department of Physics,
    Parala Maharaja
    Engineering College
Odisha, India

Oznur Ozaltin
Department of Statistics, Faculty of
    Science, Hacettepe University
Ankara, Turkey

Ram Bilas Pachori
Department of Electrical Engineering,
    Indian Institute of Technology Indore
Indore, India

Saurabh Pal
University of Calcutta
Kolkata, India

Srikireddy Dhanunjay Reddy
Department of Electronics and
    Communication Engineering, Indian
    Institute of Technology
Roorkee, India

Mohammad Ebad Ur Rehman
Department of Medicine, Rawalpindi
    Medical University
Punjab, Pakistan

Ramnivas Sharma
Department of Electrical Engineering,
    Malaviya National Institute of
    Technology
Jaipur, India

Vivek Kumar Singh
Department of Electrical Engineering,
    Indian Institute of Technology
Indore, India

Dilip Singh Sisodia
Department of Computer Science &
    Engineering, National Institute of
    Technology
Raipur, India

Abdulhamit Subasi
Department of Computer Science,
    College of Engineering, Effat
    University
Jeddah, Saudi Arabia

Tuba Nur Subasi
Faculty of Medicine, Trakya
    University
Edirne, Turkey.

Gary Tse
School of Nursing and Health
    Studies, Hong Kong Metropolitan
    University
Hong Kong, China

Rajesh Kumar Tripathy
Department of Electrical and
    Electronics Engineering,
    BITS-Pilani
Hyderabad, India

Siran Wang
Chu Kochen Honors College, Zhejiang
    University
Hangzhou, China

1 Introduction to EEG Signal Recording and Processing

Kritiprasanna Das, Vivek Kumar Singh, and Ram Bilas Pachori

1.1 INTRODUCTION

A German neuropsychiatrist, Hans Berger, discovered electroencephalography (EEG) for humans (1). EEG is an electrophysiological method for capturing electrical activity generated by a group of neural populations in the human brain. Due to the exceptional temporal sensitivity of EEG, it is useful for studying dynamic brain activity. EEG is especially helpful for diagnosing patients with epilepsy and probable seizures, unusual spells, dementia, etc. It has been extensively used for research in the areas of neuroscience, cognitive psychology, cognitive science, and neurolinguistics (2).

EEG has also been used for a number of other clinical purposes. It can be used to track the level of anesthesia during surgery, to detect motor-imagery movements, etc. because it is so sensitive to detecting quick changes in brain activity. EEG has been shown to be very useful for monitoring the depth of anesthesia and for keeping an eye out for prospective issues like ischemia or infarction. The average of EEG waveforms corresponding to a particular task gives rise to evoked potentials (EPs) and event-related potentials (ERPs). These potentials represent the neural activity of interest that is temporally related to a specific stimulus. In both clinical practice and research, EPs and ERPs are utilized to examine auditory, visual, somatosensory, and higher cognitive functioning.

In the cerebral cortex, the cortical pyramidal neurons, positioned perpendicular to the surface of the brain, are assumed to be the main source of EEG signals, which essentially are the excitatory and inhibitory postsynaptic potentials of relatively large groups of synchronously firing neurons. Traditional EEG recorded on the scalp or cortical surface cannot record the momentary local field potential changes resulting from neuronal action potentials (3, 4).

1.2 EEG ACQUISITION

Acquisition of physiological signals and images has become vital in the early diagnosis of various diseases. Some of the recordings of the electrical activity of the human body include electrocardiogram (ECG), electromyogram (EMG), EEG, electrogastrogram (EGG), and electrooculogram (EOG) signals, which represent the electrical activity of the heart, muscles, brain, stomach, and eye, respectively. Similarly, magnetoencephalography (MEG) is the measurement of magnetic field generated due to electrical activity in the neurons of the human brain.

Various imaging techniques play an equal role in the early or on-time diagnosis of disease such as sonography (ultrasound imaging), magnetic resonance imaging (MRI), functional MRI (fMRI), computerized tomography, positron emission tomography, single photon emission tomography, and near-infrared spectroscopy. EEG, MEG, and fMRI signals and images capture the physiological and functional changes happening inside the brain. The applications of fMRI compared with those of EEG or MEG signals are limited because of the following reasons (5):

- fMRI has very low time resolution, approximately 2 frames/sec.
- fMRI cannot capture various mental activities and brain disorders as they have less effect on the level of blood oxygenation.
- fMRI is limited in access as well as costly.
- Additionally, fMRI demands a sophisticated lab setup.

In contrast with other neuroimaging techniques, EEG does not have these limitations. Here, we briefly evaluate EEG technology. The very first electrical neural activity was captured with the help of a simple galvanometer. As the pointer variation of the galvanometer was very fine, light was projected on the galvanometer and reflected on a wall with the help of a mirror in order to record or visualize the variations. Lippmann and Marey introduced the capillary electrometer, and in 1903, Einthoven introduced the string galvanometer, a very sensitive and accurate measuring instrument that enabled photographic recording and became a standard instrument for a few decades. The recent EEG recording systems consist of delicate electrodes, one differential amplifier per channel or electrode, filters, and registers, and the captured multichannel EEG signals are plotted on paper. After the arrival of this product in the market, researchers saw the need for a system that could digitize (using multichannel analog-to-digital converters) and store the plotted data for analysis on computers (5).

The computerized EEG recording systems are equipped with stimulations, controls of the sampling frequency, and some advanced signal processing tools to preprocess the recorded signals. Generally, most of the significant information is present in the 0–100 Hz frequency region; therefore, a minimum sampling rate of 200 samples/sec is required. There are few applications of EEG signal processing for which high-frequency information is important; hence, EEG recording devices allow for choosing samplings rate of up to approximately 2000 samples/sec. For the quantization of the EEG signals, 16-bit quantization is very popular as it maintains the diagnostic information. This makes the archiving volume of the EEG signals very high for applications like epileptic seizure monitoring and sleep EEG records, and archiving large volumes of longer-duration EEG signals requires large storage facilities at diagnostic or research centers and hospitals.

The recording electrodes and their proper functioning play a very important role in the quality of acquired data. There are several types of recording electrodes used for EEG signal recording, namely, pregelled and gel-less disposable electrodes; tin, stainless steel, silver, or gold reusable disc electrodes; saline-based electrodes; electrode caps; needle or cortical electrodes, etc. Ag-AgCl disk electrodes are the most commonly used, with a diameter of less than 3 mm and wired leads that can be connected to amplifiers. The cortical electrodes are used for recording invasive EEG signals by implanting them under the skull via minimally invasive operations.

The use of high-impedance electrodes or the presence of high impedance between the cortex and the electrodes can lead to the severe distortion of EEG signals. Recording EEG signals with electrodes with an impedance less than 5000 Ω provides satisfactory signal quality. Due to very low amplitude (in the range of microvolts), a high gain amplifier is required; a typical EEG amplifier usually provides a voltage gain of 5000 to 50000. Distribution of the potentials is nonuniform over the scalp because of the spiral and layered structure of the brain, which can affect the results of the source localization performed using EEG signals.

1.2.1 CONVENTIONAL EEG ELECTRODE POSITIONING

Figure 1.1 depicts a conventional 10–20 electrode positioning that is recommended by the International Federation of Societies for Electroencephalography and Clinical Neurophysiology. It consists of a total of 21 electrodes, including the two earlobe electrodes (A1 and A2) that are used as reference. The name of the

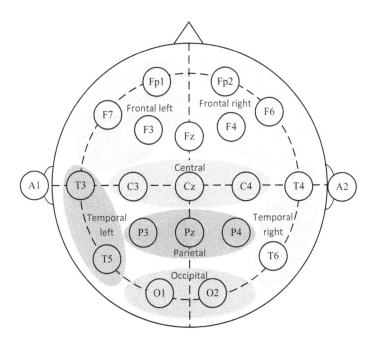

FIGURE 1.1 10–20 international standard for EEG electrode position.

electrodes is given based on the cerebral lobe position, e.g., an electrode placed on the frontal lobe is labeled "F." The electrodes in the left hemisphere are numbered with odd numbers, the right hemisphere electrodes are numbered with even numbers, and the electrodes on the longitudinal fissure are marked with a letter "z," like "Cz."

To record EEG with higher spatial resolution, a large number of electrodes are required positioned equidistantly in between the electrodes in a typical 10–20 system; for example, F2 is placed between F4 and Fz. Extra electrodes are sometimes employed to measure the ECG, EOG, and EMG of the eyelid and surrounding muscles, which may help in artifact removal. Additionally, for some applications like brain–computer interface (BCI), fewer electrodes and even just a single electrode are used.

The EEG signal can be recorded in a bipolar (differential) or unipolar (referential) fashion. In bipolar recording, two inputs of the amplifier are attached, with two EEG electrodes placed in different locations on the scalp. Bipolar recordings are suitable for the analysis of localized neural activity. On the other hand, in unipolar recording, one or two reference electrodes are commonly connected to one input of the amplifier, and the other input of the amplifier is connected to the general EEG electrodes (e.g., F1, F2). In literature, the reference electrodes have been placed in different locations such as Cz, the earlobe, and the mastoid. There are also reference-free EEG acquisition approaches that employ a common

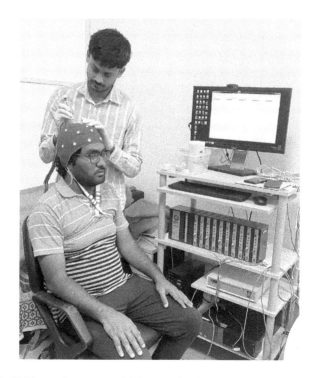

FIGURE 1.2   EEG cap placement and filling conductive gel.

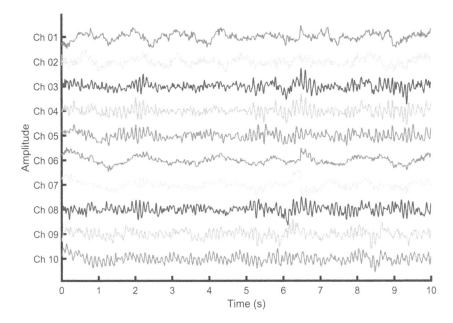

FIGURE 1.3 EEG signals (Ch i represents the ith channel EEG signal; here, i varies from 1 to 10).

average reference. In a typical EEG recording experiment, the following steps are performed (6):

(1) A technician measures your head and traces your scalp with a special pencil to indicate where the electrodes will be attached.
(2) Electrodes are attached using adhesive conductive paste. Sometimes, an elastic cap with electrodes inside is used to place the electrodes all over the scalp easily.
(3) A proper connection is established between the electrodes and the amplifier for recording the EEG.

A typical experimental procedure for recording EEG signals is shown in Figure 1.2, where an EEG cap is placed on the subject's head and the technician fills the conductive electrode gel to establish a proper connection with the scalp. Figure 1.3 shows a 10-channel differentially recorded EEG signal (7).

1.2.2 EEG ACQUISITION DEVICE

There are numerous commercially accessible EEG recording devices. Based on the application requirement, properly choosing an EEG device is important. These devices can be categorized based on the connectivity with the computer system, electrode connection, etc. (8). Several parameters such as number of channels, device and electrode connectivity, and amplifier gain need to be carefully considered in the selection of EEG devices.

1.2.3   WIRED AND WIRELESS EEG

The connectivity between the acquisition devices is established using wired technology or wirelessly using Bluetooth or WiFi. Wired EEG devices provide more stable data transfer with a higher data transfer rate. Wireless EEG devices offer freedom of movement, but data loss can happen when wireless connectivity is lost. With wired connections, movements of cables and electrodes introduce artifacts in both devices.

1.2.4   ELECTRODE CONNECTION

The proper connection between the electrode and the scalp is of utmost necessity to obtain a good-quality signal. To establish the connection between the electrode and scalp, conductive gel, saline solution, or conductive adhesive paste is used, which reduces the impedance. A few modern EEG devices also use dry electrodes. For short-duration experiments, saline solution-based or dry electrodes are suitable as the setup time for these kinds of electrodes is shorter. However, with time, saline water will dry, and the impedance between the scalp and electrodes will increase, which deteriorates the signal quality. Due to this, conductive gel-based electrodes are preferred for long-duration experiments.

1.2.5   WEARABLE EEG

For a few applications like human–computer interaction, imagined speech recognition-based BCI systems, EEG-based rehabilitation devices, epileptic seizure onset prediction, continuous recording and monitoring for months are necessary, but placing electrodes with wires and other bulky accessories reduces the user's comfort and restricts the long-term recording of EEG signals. Recently, researchers have been trying to develop wearable EEG electrodes with reduced channels (9, 10). A flexible electronic system printed on the scalp, like a tattoo or ear electrode, was developed for recording EEG signals for BCI applications (11).

1.3   ARTIFACTS

Artifacts are undesired signals that adversely affect the signal of interest. It is desirable to prevent artifacts from appearing while recording, but EEG signals are, unfortunately, frequently corrupted by physiological and environmental factors other than cerebral activity. An important component of EEG signal processing is removing noise and artifacts, which is typically required for more trustworthy signal analysis. The two main types of artifacts are physiological/biological caused by noncerebral physiological sources and nonphysiological artifacts caused by electrical phenomena or equipment in the recording environment. Physiological artifacts include eye, cardiac, glossokinetic, respiratory, pulse, sweat, and muscle movement artifacts (2). Powerline noise, cable movement, and electromagnetic interferences are common environmental artifacts. In the next section, we will give a brief overview of artifacts common to EEG.

1.3.1 PHYSIOLOGICAL ARTIFACTS

1.3.1.1 Ocular Artifacts

Eye movements and blinks during EEG recording are the causes of ocular artifacts. To be more precise, retinal and corneal dipole orientation alterations cause eye movement artifacts, and alterations of contact of the cornea with the eyelid affect ocular conductance, which results in blink artifacts. Moreover, ocular artifacts can spread to the head's surface and be recorded by the EEG electrodes as a result of the volume conduction effect. EOG often has a frequency similar to those of EEG signals and an amplitude many times greater than those of EEG, which disqualifies frequency filtering as an artifact removal technique (12).

1.3.1.2 Muscle Artifacts

Activities from different groups of muscles contaminate EEG, which is known as muscle artifacts. These artifacts can be caused by the subject's talking, sniffing, swallowing, or muscle contraction and stretch close to the signal recording sites. Depending on the muscles that are stretched and contracted, the amplitude and shape of the EMG will change, but muscle activities detected by EMG have a wide frequency range between 0 Hz and 200 Hz. Obtaining the activity from a single channel measurement is extremely difficult compared with EOG and eye tracking. As a result, it can be extremely difficult to eliminate EMG artifacts. Significant statistical separation exists between EMG contamination and EEG in both time and space. This suggests that using independent component analysis to exclude EMG contamination would be a good idea.

1.3.1.3 Cardiac Artifacts

Positioning EEG electrodes on or close to a blood vessel can create cardiac artifacts due to the expansion and contraction caused by the heart. It is challenging to eliminate these pulse distortions since they can appear in the EEG with similar waveforms and with a frequency of about 1.2 Hz (13). The electrical activity of the heart, ECG, can also be contaminated with EEG. As ECG can be monitored with a recognizable regular pattern and recorded separately from brain activity, unlike pulse artifacts, it may be simpler to remove these artifacts by simply utilizing a reference waveform.

1.3.2 EXTRINSIC ARTIFACTS

In addition to the aforementioned artifacts, EEG measurement is negatively impacted by external sources of artifacts including cable movements, misplacement of electrodes, etc. Proper planning and improved signal acquisition can be helpful in minimizing these kinds of artifacts. Another external artifact that influences EEG data is electromagnetic interference from the environment which belongs to a specific frequency band. Power line artifacts are generated due to interference from power sources with a frequency of 50/60 Hz (14). Movement of any part of recording devices, like electrode wires, can generate artifacts.

1.4   SIGNAL PROCESSING FOR EEG ANALYSIS

EEG signals present a landscape of neural activity in the brain with high temporal resolution. At the same time, EEG suffers from drawbacks like proneness to noise prone and high complexity. Due to very low amplitudes (in the microvolt range), various physiological and environmental noises easily affect the EEG signal by degrading the signal-to-noise ratio. Analysis of such noisy EEG signals may lead to erroneous interpretation.

A few of these artifacts can be reduced by taking proper measures during recording, but most of them are unavoidable. To improve the signal quality and make it eligible for further processing, artifact removal is a useful pre-processing technique for which signal processing has been proven to be a valuable tool. Signal processing is also necessary for extracting useful information from complex EEG signals for meaningful interpretation. Many adaptive signal decomposition techniques like multivariate Fourier–Bessel empirical mode decomposition (15), multivariate iterative filtering (16), and sparse spectrum-based swarm decomposition (17) have been used for decomposing the EEG signal and feature extraction.

Time–frequency representation of EEG signals is also helpful for classification with deep neural network (18–20). Additionally, instead of using deep learning techniques like convolutional neural network (CNN) for classification, activation, or output from a particular layer or multiple layers are used together as the deep features, which can be further classified with the help of machine learning classifiers (17). A general approach for the automated classification of EEG signals based on signal processing and artificial intelligence incorporates signal recording, preprocessing, signal decomposition and feature extraction, and classification, which is shown in Figure 1.4.

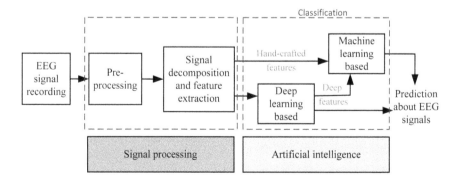

FIGURE 1.4   A block diagram of a general approach for the automated classification of EEG signals.

1.4.1 ARTIFACT REMOVAL

EEG signals are employed as a cutting-edge diagnostic tool for various neural illnesses, in BCI applications, and in studying fundamental neuroscience because they capture electrical activities produced by brain cells. But often, undesirable artifacts taint the EEG signals and make it difficult to interpret the neural activity (21–23). Signal processing techniques have been proposed for effectively removing artifacts.

A state-of-the-art technique was proposed for removing ocular artifacts, frequency-spatial filtering (21) using empirical wavelet transform (EWT) and dictionary-based spatial filtering. The authors built an isolated artifact dictionary by selecting the contaminated EEG channels and filtering the EWT-based frequency. More preciously, the authors removed the delta rhythms of the highly contaminated channels and added them to an artifact dictionary. Afterward, the ocular artifact was isolated by spatially filtering the delta rhythms of the multichannel EEG data using the developed dictionary. After they eliminated the artifact components, the authors reconstructed a clean EEG delta rhythm using inverse spatial filtering. In the end, to obtain ocular-artifact-free signals, they merged the clear delta rhythms with other EEG rhythms. The suggested technique eliminated the ocular artifacts while leaving the baseline EEG data unchanged.

1.4.2 EEG RHYTHM SEPARATION

Visual examination of EEG data can be used to diagnose various brain illnesses. EEG signals typically have a frequency range of 0.1 to 100 Hz, and based on the frequency, they can be further divided into five distinct rhythms: delta (0.1–4 Hz), theta (4–8 Hz), alpha (8–13 Hz), beta (13–30 Hz), and gamma (30–100 Hz). Clinical professionals with expertise in this area are familiar with the manifestation of brain rhythms in EEG signals. The amplitudes and frequency of these rhythms vary depending on the human's state, such as awake or asleep. Age also alters the properties of the rhythm waves.

Various techniques have been proposed for obtaining these rhythms from EEG signals. Band-pass filters with predefined pass bands have commonly been used to extract the EEG rhythms. Several studies have shown that instead of using any arbitrary pass band, adaptively choosing oscillatory components can give more useful information (16, 24). Motivated by this, adaptive decomposition techniques have been used for rhythm separation, where the signal has been decomposed into intrinsic mode functions (IMFs), and based on the instantaneous frequency of the IMF, groping is performed to obtain the EEG rhythms (25).

Multivariate EEG rhythms have been extracted from multichannel EEG signals using multivariate iterative filtering (16). The multivariate iterative filtering decomposes the EEG signal into multivariate IMFs, which are properly aligned across the different channels' oscillatory components in terms of frequency content. These IMFs are grouped based on their mean frequency to get the multivariate rhythms (16). This technique is helpful for the multivariate analysis of EEG rhythms, and the Matlab code for the same is available online on https://github.com/kpdas95/EEG_Rhythm_Separation. Figure 1.5 shows the delta, theta, alpha, beta, and gamma

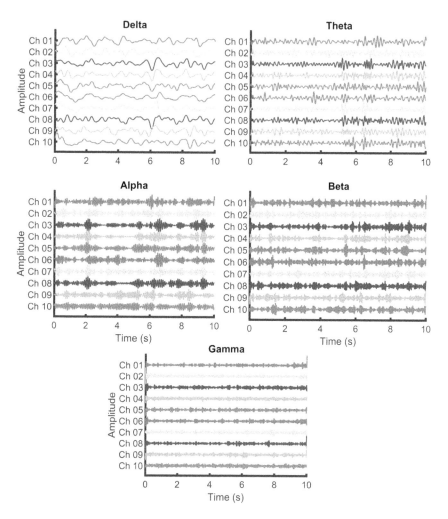

FIGURE 1.5   EEG rhythms (Ch *i* represents *i*th channel EEG signal, here *i* varies from 1 to 10).

rhythms corresponding to the EEG signal shown in Figure 1.3, extracted using multivariate iterative filtering-based rhythm separation (16).

1.4.3 FEATURE EXTRACTION

Due to the complexity of EEG signals, it is highly challenging to extract information from them using the naked eye. These days, we use sophisticated automatic processing methods to retrieve hidden information from EEG data owing to computers. There exist various ways to represent EEG using features from domains such as time (mean, standard deviation, entropy (16, 26)), frequency (mean frequency, bandpower), time and frequency together (Shannon entropy, time-varying energy, instantaneous amplitude, and frequency (27, 28)), and synchronicity, which involves the relationships between two or more EEG channels (coherence, mutual information, correlation), merely to name a few. Deep learning networks also have been used to extract automated or deep features from EEG signals.

1.5 APPLICATIONS

A large number of studies have demonstrated that EEG is useful for assessing human mental health states, clinical conditions, imagination, thoughts, etc. EEG finds applications in areas like clinical diagnosis, BCI, biometrics, fundamental neuroscience, neuromarketing, and custom solutions (4, 8, 29).

1.5.1 CLINICAL APPLICATIONS

EEG is a very useful tool for various disease diagnoses and predictions, including but not limited to epilepsy, dyslexia, Alzheimer's disease, Parkinson's disease, attention deficit hyperactivity disorder (ADHD), sleep disorders, Huntington's disease, anxiety and depression, and schizophrenia. Monitoring EEG signals during neurosurgery helps to complete the procedure smoothly and properly and increases the success rate. We next describe a few representative clinical diagnostic applications.

1.5.1.1 Epilepsy Detection

The simultaneous irregular firing of a neuronal population causes epilepsy, which is the second most well-known neurological condition in the brain. Almost 60 million people worldwide are affected by epilepsy. EEG is a gold-standard diagnostic tool for epilepsy, but it requires long monitoring of the signal; however, it is laborious and time-consuming to manually monitor the patient's EEG signal for an extended period. Furthermore, muscle artifacts, background noise, and neurological symptomatology may contaminate the recorded EEG data. Hence, a system that automatically detects epileptic seizures will make it easier to monitor and treat them in real time (15, 20, 30–33). Madhavan et al. (33) proposed automated classification of focal and nonfocal EEG signals. The nonstationary EEG signal is represented in the time–frequency plane using synchro-squeezing transform and wavelet SST. Two-dimensional CNN is used to classify the time–frequency matrix of EEG signals into focal and nonfocal classes.

1.5.1.2 Schizophrenia Detection

Schizophrenia is a severe and chronic mental illness that affects 20 million individuals globally, or roughly 1% of the total population (34), with more than one million cases occurring in India. In active states of schizophrenia, it shows symptoms like hallucinations (experience of a sense that does not actually occur, such as sight, sound, taste, or smell), delusions (fixed false belief), abnormal motor behavior (movements that can range from childish silliness to unpredictable agitation or purposeless movements), and disorganized speech and thinking (35). Despite the fact that there is no permanent cure for schizophrenia, the majority of schizophrenia symptoms can be considerably improved by proper treatment, and the risk of recurrence can be reduced.

Schizophrenia mimics many symptoms of other neurological disorders, which can mislead, and substance misuse can impede diagnosis and therapy. A psychiatrist needs to conduct a thorough examination to accurately diagnose schizophrenia, which is a lengthy, time-consuming procedure that can introduce subjective error. The likelihood of substance abuse can be decreased, and therapy can begin right away with accurate schizophrenia predictions. Ninety percent of schizophrenia patients reside in low- and middle-income countries, and more than 69% of them do not receive the proper care (34). An important problem is the lack of access to mental health services. In this case, simple, effective, and affordable detection techniques might be advantageous for all of humanity.

An approach based on multivariate iterative filtering was developed for the reliable prediction of schizophrenia from multichannel EEG signals. The signal data were decomposed into multivariate oscillatory modes using iterative filtering, and the modes were used to obtain EEG rhythms, which were represented using Hjorth parameters features. These features were ranked based on Student t tests, and significant features were selected. Machine learning classifiers were then developed for classifying the EEG signals using the features (16).

1.5.1.3 Sleep Analysis

In sleep apnea, airflow is temporarily stopped or reduced during sleep for a few seconds. This decrease in breathing is accompanied by loud snoring, which could cause the person to feel choked and awaken. Researchers have developed a number of techniques to diagnose sleep apnea, which polysomnography (PSG) indicated as likely the most effective; PSG monitors several physiological parameters like brain waves, heart rate, breathing pattern, eye movements, blood oxygen level, limb and body movements, and the snoring sound. However, the simultaneous recording of these parameters is complicated and creates user discomfort, and the analysis of PSG is cumbersome and tedious.

There are six stages of sleep for a healthy person: awake, S1, S2, S3, S4, and rapid eye movement (18). Precise sleep stage grading can provide clinical information for identifying people with sleep disorders (36). The automatic detection of sleep stages and sleep apnea from biological signals using signal processing and artificial intelligence-based techniques has been reported in many studies (17, 37). In one such study, the nonstationary signal analysis technique Fourier-Bessel decomposition (FBD) and deep learning classifier, are used for scoring the different sleep stages from the EEG signal (18). FBD was used to decompose EEG signals into oscillatory modes or Fourier–Bessel intrinsic band functions, which are suitable for obtaining instantaneous frequency and amplitude envelope using Hilbert transform. The authors obtained the time–frequency image from the instantaneous frequency and amplitude envelope of the EEG signal and used CNN to classify the image. The authors used their developed method and EEG signals to categorize six different stages of sleep.

1.5.1.4 Parkinson's Disease Detection

Parkinson's disease is a chronic progressive neurodegenerative disorder that affects movements (38). It predominantly affects a specific area of the brain called substantia nigra where dopamine-producing (dopaminergic) neurons are located. Parkinson's disease signs and symptoms include tremors, slowed movement (bradykinesia), rigid muscles, impaired posture and balance, loss of automatic movements, speech changes, urinary disturbances, and difficulties in writing, but they differ from person to person, which makes early diagnosis challenging. Additionally, in the early stages of Parkinson's disease, behavioral symptoms are barely noticeable and hard to diagnose.

To date, there are no definitive imaging or biological markers for detecting Parkinson's disease (39). A well-trained neurologist can diagnose Parkinson's disease based on medical history, signs and symptoms, and physical and neurological examinations. Automated disease diagnosis techniques can be helpful for early diagnosis and starting the treatment (40). For instance, researchers used the change in EEG dynamics assessed through chaos theory-based embedding reconstruction as a biomarker to detect Parkinson's disease (41), and the synchronization of the higher-frequency components' amplitude and beta rhythms' phase increased significantly. Based on this information, the authors developed a deep neural network for classifying EEG signals corresponding to Parkinson's disease.

1.5.2 BRAIN–COMPUTER INTERFACE AND REHABILITATION APPLICATIONS

In many pieces of research, EEG signals have been used in BCI-based rehabilitation applications to establish alternative communication paths between the brain and the outside world.

1.5.2.1 Motor Imagery BCI

BCI helps in establishing alternative communication paths in many ways, including decoding imagery motor movement from the brain signal itself so that it can be used as a command for the external world. Motor imagery BCI is very useful for rehabilitation after stroke or trauma and for developing supportive and prosthetic devices (13, 42). Common spatial patterns (CSP) are used to extract features from EEG signals (42, 43); the operating frequency bands–which are often manually chosen or set to a wide frequency range—determine the performance of CSP. However, these approaches function poorly because the frequency band varies from subject to subject or even trial to trial for different motor imagery tasks.

In one motor imagery approach, multivariate iterative filtering was used to handle data variability. Multivariate iterative filtering adaptively decomposed the multichannel EEG data and helped with automatically selecting the optimum frequency band. Then, from each band, CSP features were extracted and classified using a linear discriminant analysis-based classifier (24).

1.5.2.2 Emotion Recognition

Emotions are essential to human existence and have an impact on daily functions like cognition, decision-making, and intelligence, among many others; a recent trend in the field of human–computer interactions is emotional artificial intelligence. Additionally, emotion has a direct connection to many mental diseases, including depression, ADHD, autism, and gaming addiction. The importance of understanding emotion gives birth to a new scientific field, affective computing, which primarily deals with identifying and modeling human emotions. Compared with other methods that rely on outward manifestations like facial expressions, gestures, or speech signals (44), which may show faked emotions, EEG signals are found to be more compelling for emotion recognition (26, 44–46).

Researchers used multivariate Fourier–Bessel series expansion-based empirical wavelet transform to detect emotion by decomposing multichannel EEG signals into narrow-band subband signals. They selected different successive joint instantaneous amplitude and frequency of subband signals to have multiscaling properties in the spectral domain, added subband signals, and computed the entropy of the cumulative signals as temporal multiscale entropies. They then smoothed the spectral and temporal multiscale entropies and classified them using an autoencoder-based random forest classifier for emotion classification (45).

1.5.2.3 Cognitive Workload Assessment

Studying the mental effort involved in problem-solving is crucial for fully comprehending how the brain allocates cognitive resources to interpret information. Mental effort reflects the quantity of cognitive resources allocated for a particular task. EEG is an effective physiological signal-based approach for assessing mental workload.

Researchers induce mental effort by giving scientific problems of differing levels of complexity and comparing the power in different EEG rhythms during problem-solving is compared with reference intervals where the subject was not performing any task. The authors calculated the percentage changes in rhythm power, and a positive value or increase indicated event-related synchronization; a negative value or decrease indicated event-related resynchronization. The authors found increased alpha (lower: 8–10 Hz) desynchronization in the occipital and parietal regions and theta (4–7 Hz) synchronization in the frontal lobe. These findings suggest that mental effort due to scientific problem solving demands cognitive resources like working memory and visuospatial and semantic processing (47).

1.5.2.4 Imagined Speech

Several diseases, for example, pseudocoma or lock-in syndrome, affect the speech generation process, causing subjects to lose their ability to communicate verbally. In many cases, the brain or central nervous system of such patients works normally; therefore, the BCI can be a substitute for reading the commands from the brain itself. Many researchers try to decode the EEG signals corresponding to imagined speech. The imagined-speech-based BCI system is useful for a person with a speech disorder not in the central nervous system (48, 49).

Researchers proposed multiscale signal decomposition to classify EEG signals corresponding to five vowels for an imagined speech BCI system. They used multivariate fast empirical mode decomposition to decompose multichannel EEG signals into oscillatory components at different scales. They computed statistical features like slope domain entropy, sample entropy, bubble entropy, and energy from the oscillatory modes and used gradient boosting-based machine learning algorithms to classify the EEG signals (48).

1.5.3 FUNDAMENTAL NEUROSCIENCE

EEG signals have been used to understand the complex neural mechanisms underlying different cognitive processes, brain function, and dysfunction. Here, we present a few areas of neuroscience in which EEG signals have been used to understand brain functioning.

1.5.3.1 Visual Object Recognition

Computational neuroscientists try to reveal the brain's neural functioning behind visual processing for object recognition using computational and mathematical models. Through a number of phases of linear and nonlinear transformations functioning at a millisecond timeframe, the human brain recognizes visual objects. Signal processing and machine learning methods have been used to explain and predict these transformations.

The authors of (50) used a large dataset of EEG signals during the visualization of 16,740 image conditions to model visual object recognition; they performed a total of 82,160 trials for the image conditions. Based on this dataset, they developed visual cognition models for predicting synthesized EEG signals as a response to an image, identifying the image conditions from synthesized EEG data. Their results reflect the testing of multiple trials and image conditions.

1.5.3.2 Study of Visual Imagery and Perception

Visual imagery and perception share similar brain resources, which has been shown by studying the EEG signals during visual imagery and perception (51, 52). These help biological organisms in cognizing beyond their immediate response to a physically presented stimulus to behave adaptively and with enough flexibility.

Multivariate pattern analysis of EEG signal using signal processing and machine learning algorithms has shown that alpha rhythm in the parieto-occipital cortex shares representation for visual imagery and perception (52). The study was performed on EEG signals recorded from 38 subjects during visual perception and visual imagery tasks. For the visual imagery task, the name of the object to be imagined was uttered to trigger the subject's imagination.

1.5.3.3 Effect of Meditation on Brain

An exponentially growing number of researchers are searching for the biological mechanism underlying the beneficial effects of meditation. There exist several pieces of evidence supporting its positive impacts on both physical and mental health. We briefly discuss a study in which the authors examined the effect of meditation on the human brain with the help of EEG signal analysis. In (7), EEG rhythm powers were used as a marker to analyze the effect of mantra meditation (with the "Hare Krishna" mantra) on the human brain. The authors computed the EEG rhythm powers using Fourier–Bessel series expansion before and after the mantra meditation. The alpha band power increased significantly after meditation, indicating a calm and relaxed state of mind.

1.6 CONCLUSION

Electroencephalography (EEG) has been widely accepted and used as one of the most common neuroimaging techniques due to its multiple advantages such as high temporal resolution, lower cost of recording, and direct measurement of electrical activity at the neural population level. In this chapter, we provided a brief overview of EEG technologies, recording procedures, sample waveforms, processing methods, and applications. We discussed signal processing and machine learning-based approaches for disease diagnosis, brain–computer interfaces, and fundamental neuroscience research to help build an overview of the applicability of EEG.

REFERENCES

1. Donald L Schomer and Fernando Lopes Da Silva. *Niedermeyer's electroencephalography: Basic principles, clinical applications, and related fields.* Lippincott Williams & Wilkins, 2012.
2. Jose Antonio Urigüen and Begoña Garcia-Zapirain. EEG artifact removal state-of-the-art and guidelines. *Journal of Neural Engineering*, 12(3):031001, 2015.
3. Michael X Cohen. Where does EEG come from and what does it mean? *Trends in Neurosciences*, 40(4):208–218, 2017.
4. Mike X Cohen. *Analyzing neural time series data: Theory and practice.* MIT Press, 2014.
5. Saeid Sanei. *Adaptive processing of brain signals.* John Wiley & Sons, 2013.
6. EEG (electroencephalogram). URL: https://www.mayoclinic.org/tests-procedures/eeg/about/pac-20393875.
7. Kritiprasanna Das, Pankaj Verma, and Ram Bilas Pachori. Assessment of chanting effects using EEG signals. In *2022 24th International Conference on Digital Signal Processing and its Applications (DSPA)*, pages 1–5. IEEE, 2022.
8. Mahsa Soufineyestani, Dale Dowling, and Arshia Khan. Electroencephalography (EEG) technology applications and available devices. *Applied Sciences*, 10(21):7453, 2020.
9. Mitchell A Frankel, Mark J Lehmkuhle, Mark C Spitz, Blake J Newman, Sindhu V Richards, and Amir M Arain. Wearable reduced-channel EEG system for remote seizure monitoring. *Frontiers in Neurology*, 1842, 2021.
10. Joo Hwan Shin, Junmo Kwon, Jong Uk Kim, Hyewon Ryu, Jehyung Ok, S Joon Kwon, Hyunjin Park, and Tae-il Kim. Wearable EEG electronics for a brain–AI closed-loop system to enhance autonomous machine decision-making. *npj Flexible Electronics*, 6(1):32, 2022.
11. Musa Mahmood, Deogratias Mzurikwao, Yun-Soung Kim, Yongkuk Lee, Saswat Mishra, Robert Herbert, Audrey Duarte, Chee Siang Ang, and Woon-Hong Yeo. Fully portable and wireless universal brain–machine interfaces enabled by flexible scalp electronics and deep learning algorithm. *Nature Machine Intelligence*, 1(9):412–422, 2019.
12. Xiao Jiang, Gui-Bin Bian, and Zean Tian. Removal of artifacts from EEG signals: A review. *Sensors*, 19(5):987, 2019.
13. Pramod Gaur, Anirban Chowdhury, Karl McCreadie, Ram Bilas Pachori, and Hui Wang. Logistic regression with tangent space based cross-subject learning for enhancing motor imagery classification. *IEEE Transactions on Cognitive and Developmental Systems*, 14(3):1188–1197, 2022.
14. Alain de Cheveigné. ZapLine: A simple and effective method to remove power line artifacts. *NeuroImage*, 207:116356, 2020.

15. Abhijit Bhattacharyya and Ram Bilas Pachori. A multivariate approach for patient-specific EEG seizure detection using empirical wavelet transform. *IEEE Transactions on Biomedical Engineering*, 64(9):2003–2015, 2017.
16. Kritiprasanna Das and Ram Bilas Pachori. Schizophrenia detection technique using multivariate iterative filtering and multichannel EEG signals. *Biomedical Signal Processing and Control*, 67:102525, 2021.
17. Shailesh Vitthalrao Bhalerao and Ram Bilas Pachori. Sparse spectrum based swarm decomposition for robust nonstationary signal analysis with application to sleep apnea detection from EEG. *Biomedical Signal Processing and Control*, 77:103792, 2022.
18. Vipin Gupta and Ram Bilas Pachori. FBDM based time-frequency representation for sleep stages classification using EEG signals. *Biomedical Signal Processing and Control*, 64:102265, 2021.
19. Ram Bilas Pachori. *Time-frequency analysis techniques and their applications*. CRC Press, 2023.
20. Rishi Raj Sharma and Ram Bilas Pachori. Time–frequency representation using IEVDHM–HT with application to classification of epileptic EEG signals. *IET Science, Measurement & Technology*, 12(1):72–82, 2018.
21. Abhijit Bhattacharyya, Aarushi Verma, Radu Ranta, and Ram Bilas Pachori. Ocular artifacts elimination from multivariate EEG signal using frequency-spatial filtering. *IEEE Transactions on Cognitive and Developmental Systems*, 15(3):1547–1559, 2023.
22. Pranjali Gajbhiye, Rajesh Kumar Tripathy, Abhijit Bhattacharyya, and Ram Bilas Pachori. Novel approaches for the removal of motion artifact from EEG recordings. *IEEE Sensors Journal*, 19(22):10600–10608, 2019.
23. Pranjali Gajbhiye, Rajesh Kumar Tripathy, and Ram Bilas Pachori. Elimination of ocular artifacts from single channel EEG signals using FBSE-EWT based rhythms. *IEEE Sensors Journal*, 20(7):3687–3696, 2019.
24. Kritiprasanna Das and Ram Bilas Pachori. Electroencephalogram based motor imagery brain computer interface using multivariate iterative filtering and spatial filtering. *IEEE Transactions on Cognitive and Developmental Systems*, 15(3):1408–1418, 2023.
25. Varun Bajaj and Ram Bilas Pachori. Separation of rhythms of EEG signals based on Hilbert-Huang transformation with application to seizure detection. In *Convergence and Hybrid Information Technology: 6th International Conference, ICHIT 2012, Daejeon, Korea, August 23–25, 2012. Proceedings 6*, pages 493–500. Springer, 2012.
26. Aditya Nalwaya, Kritiprasanna Das, and Ram Bilas Pachori. Automated emotion identification using Fourier–Bessel domain-based entropies. *Entropy*, 24(10):1322, 2022.
27. Vivek Kumar Singh and Ram Bilas Pachori. Sliding eigenvalue decomposition for non-stationary signal analysis. In *2020 International Conference on Signal Processing and Communications (SPCOM)*, pages 1–5. IEEE, 2020.
28. Vivek Kumar Singh and Ram Bilas Pachori. Iterative eigenvalue decomposition of Hankel matrix: An EMD like tool. *TechRxiv*, 2022, DOI: https://doi.org/10.36227/techrxiv.21730487.v1.
29. Sai Pranavi Kamaraju, Kritiprasanna Das, and Ram Bilas Pachori. EEG based biometric authentication system using multivariate FBSE entropy. *TechRxiv*, 2023, DOI: https://doi.org/10.36227/techrxiv.23244209.v1.
30. Arti Anuragi, Dilip Singh Sisodia, and Ram Bilas Pachori. Automated FBSE-EWT based learning framework for detection of epileptic seizures using time-segmented EEG signals. *Computers in Biology and Medicine*, 136:104708, 2021.
31. Arti Anuragi, Dilip Singh Sisodia, and Ram Bilas Pachori. Epileptic-seizure classification using phase-space representation of FBSE-EWT based EEG sub-band signals and ensemble learners. *Biomedical Signal Processing and Control*, 71:103138, 2022.

32. Sibghatullah I Khan and Ram Bilas Pachori. Empirical wavelet transform-based frame-work for diagnosis of epilepsy using EEG signals. In *AI-Enabled Smart Healthcare Using Biomedical Signals*, pages 217–239. IGI Global, 2022.

33. Srirangan Madhavan, Rajesh Kumar Tripathy, and Ram Bilas Pachori. Time-frequency domain deep convolutional neural network for the classification of focal and non-focal EEG signals. *IEEE Sensors Journal*, 20(6):3078–3086, 2019.

34. Schizophrenia. URL: https://www.who.int/news-room/fact-sheets/detail/schizophrenia.

35. What is Schizophrenia? URL: https://www.psychiatry.org/patients-families/schizophre nia/what-is-schizophrenia.

36. Reza Boostani, Foroozan Karimzadeh, and Mohammad Nami. A comparative review on sleep stage classification methods in patients and healthy individuals. *Computer Methods and Programs in Biomedicine*, 140:77–91, 2017.

37. Parisa Moridian, Afshin Shoeibi, Marjane Khodatars, Mahboobeh Jafari, Ram Bilas Pachori, Ali Khadem, Roohallah Alizadehsani, and Sai Ho Ling. Automatic diagnosis of sleep apnea from biomedical signals using artificial intelligence techniques: Methods, challenges, and future works. *Wiley Interdisciplinary Reviews: Data Mining and Knowledge Discovery*, 12(6):e1478, 2022.

38. Laurent Pezard, Robert Jech, and Evzen Ruzicka. Investigation of nonlinear properties of multichannel EEG in the early stages of Parkinson's disease. *Clinical Neurophysiology*, 112(1):38–45, 2001.

39. Parkinson's disease. URL: https://www.mayoclinic.org/diseases-conditions/parkinsons-disease/symptoms-causes/syc-20376055.

40. Ram Bilas Pachori and Kritiprasanna Das. System and method for predicting Parkinson's disease, 2022. Indian Patent, Application no: 202221027358.

41. Syed Aamir Ali Shah, Lei Zhang, and Abdul Bais. Dynamical system based compact deep hybrid network for classification of Parkinson disease related EEG signals. *Neural Networks*, 130:75–84, 2020.

42. Pramod Gaur, Harsh Gupta, Anirban Chowdhury, Karl McCreadie, Ram Bilas Pachori, and Hui Wang. A sliding window common spatial pattern for enhancing motor imagery classification in EEG-BCI. *IEEE Transactions on Instrumentation and Measurement*, 70:1–9, 2021.

43. Kai Keng Ang, Zheng Yang Chin, Haihong Zhang, and Cuntai Guan. Filter bank common spatial pattern (FBCSP) in brain–computer interface. In *2008 IEEE International Joint Conference on Neural Networks (IEEE World Congress on Computational Intelligence)*, pages 2390–2397. IEEE, 2008.

44. Arti Anuragi, Dilip Singh Sisodia, and Ram Bilas Pachori. EEG-based cross-subject emotion recognition using Fourier-Bessel series expansion-based empirical wavelet transform and NCA feature selection method. *Information Sciences*, 610:508–524, 2022.

45. Abhijit Bhattacharyya, Rajesh Kumar Tripathy, Lalit Garg, and Ram Bilas Pachori. A novel multivariate-multiscale approach for computing EEG spectral and temporal complexity for human emotion recognition. *IEEE Sensors Journal*, 21(3):3579–3591, 2020.

46. Aditya Nalwaya, Kritiprasanna Das, and Ram Bilas Pachori. Emotion identification from TQWT-based EEG rhythms. In *AI-Enabled Smart Healthcare Using Biomedical Signals*, pages 195–216. IGI Global, 2022.

47. Yanmei Zhu, Qian Wang, and Li Zhang. Study of EEG characteristics while solving scientific problems with different mental effort. *Scientific Reports*, 11(1):1–12, 2021.

48. Shaswati Dash, Rajesh Kumar Tripathy, Dinesh Kumar Dash, Ganapati Panda, and Ram Bilas Pachori. Multiscale domain gradient boosting models for the automated recognition of imagined vowels using multichannel EEG signals. *IEEE Sensors Letters*, 6(11):1–4, 2022.

49. Shaswati Dash, Rajesh Kumar Tripathy, Ganapati Panda, and Ram Bilas Pachori. Automated recognition of imagined commands from EEG signals using multivariate fast and adaptive empirical mode decomposition based method. *IEEE Sensors Letters*, 6(2):1–4, 2022.

50. Alessandro T Gifford, Kshitij Dwivedi, Gemma Roig, and Radoslaw M Cichy. A large and rich EEG dataset for modeling human visual object recognition. *NeuroImage*, 264:119754, 2022.

51. Nadine Dijkstra, Sander E Bosch, and Marcel AJ van Gerven. Shared neural mechanisms of visual perception and imagery. *Trends in Cognitive Sciences*, 23(5):423–434, 2019.

52. Siying Xie, Daniel Kaiser, and Radoslaw M Cichy. Visual imagery and perception share neural representations in the alpha frequency band. *Current Biology*, 30(13):2621–2627, 2020.

2 Artificial Intelligence-Enabled EEG Signal Processing-Based Detection of Epileptic Seizures

Abdulhamit Subasi, Muhammed Enes Subasi, and Emrah Hancer

2.1 INTRODUCTION

The World Health Organization observes that there are millions of people with epilepsy around the world (Litt & Echauz, 2002). At least 20% of all epilepsy patients still experience seizures while receiving medication because it is ineffective. This served as the motivation for various researchers to develop a novel method, chronic vagal nerve stimulation, to stop epileptic seizures (Fisher et al., 1997). Contrarily, a seizure prevention system can be used to administer medication or alert patients (Iasemidis, 2003). This seizure prevention device alerts the patient in advance of an approaching seizure, allowing them to leave risky areas like streets and swimming pools. Additionally, it allows patients to take less medication (Li & Yao, 2005). Different requirements are needed in the current setting for accurate seizure detection. The patient may not have had enough time to prepare for an approaching seizure or may not have had the awareness to know they needed to act (Blum et al., 1996).

The automatic detection mechanism used in current EEG analysis is prone to artifact-induced inaccuracy. Electroencephalographers' visual inspection occasionally yields conflicting localization conclusions. Prior to providing timely intervention, responsive neurostimulation (Carrette et al., 2015), a closed-loop autonomic therapy, requires automatic, reliable seizure onset detection and demands improvement in autonomic computing technologies via brain–computer interface (BCI).

The procedure generates a significant amount of real-time data, which creates a big data issue. For example, during the course of five days, a patient may produce 1.6 Gb of data from multiple channels with a high recording frequency (i.e., 2000 Hz) (Jayapandian et al., 2014). While changes in amplitude and frequency create difficulties for feature extraction and classification, fast real-time processing necessitates both secure storage and powerful computing power. Seizure detection techniques are

DOI: 10.1201/9781003479970-2

common, but more advancements are needed to guarantee accurate early detection (Hosseini et al., 2018).

Anticonvulsant or anti-epileptic drugs help many patients control their seizures, but about one-third of patients are unresponsive to therapy (Brodie et al., 2012). In patients with uncontrolled epilepsy, serious physical harm and death (also known as sudden unexpected death of epilepsy), are common. Therefore, precise, automated seizure detection is required to warn patients and caregivers, assist epileptologists in developing effective therapeutic approaches, and enhance the quality of life for epileptic patients (Pack, 2012).

The best indications of epileptic seizures are EEG readings. Spikes and sharp waves that appear in epileptiform EEG patterns can help identify and categorize seizures (French & Pedley, 2008). The ability to diagnose epileptic seizures with EEG signal processing has significantly improved seizure detection (Rajaei et al., 2016). Analyzing EEG signals can help us better comprehend unusual brain functions like overactive or misfiring neurons. However, due to the large volume of recorded EEG samples from numerous scalp-mounted electrodes, EEG-based seizure identification is extremely difficult (Adeli et al., 2007).

Two strategies can be used to address the issues associated with the complexity of EEG-based seizure detection. The first is to, instead of using all EEG electrodes, choose a limited number of EEG channels (typically 18–23). Since full-channel EEG monitoring is costly, time-consuming, uncomfortable, and stigmatizing, it is only useful in clinics; in addition, irrelevant channels increase noise in the feature space and reduce the accuracy of seizure detection. With recently created discrete wearable patches that may capture seizure activity in daily life utilizing individual nodes, limited-channel configurations can be constructed. These wearable technologies are now more useful and practical for daily life applications thanks to recently developed technology (Saeedi et al., 2016).

The second strategy is to convert the high-dimensional dataset into a lower dimension while maintaining the locational details of the data points. The best method for lowering the number of feature space dimensions is nonlinear data embedding, which retains distance information between neighboring data points due to the nonlinear structure of the feature space of EEG signals. To create a low-dimensional feature space for epileptic seizure detection, Birjandtalab et al. (2017) combined channel selection and dimension reduction.

For many years, the EEG has been used to clinically diagnose epilepsy. When compared with other techniques such as the electrocorticogram, EEG is a safe and reliable way to measure brain activity. It is generally known that clinical analysis of EEG traces can detect seizures. But the effectiveness of automated EEG-based approaches depends on the elements that are examined and how they are applied to categorize the signal. Epilepsy patients experience frequent seizures that cause physical or behavioral abnormalities and necessitate treatment with drugs or surgery (Alam & Bhuiyan, 2013). EEG signals are captured on numerous channels. The processing of that many channels takes some time. Due to this, EEG signal processing is crucial for speeding up the process.

Multiple learners' perspectives are combined in ensemble machine learning techniques to improve performance. Therefore, even poor classifiers (simpler learners) like decision tree approaches can still perform exceptionally well. Theoretically, ensemble classifiers perform better than single classifiers (Kuncheva, 2004; Valentini & Dietterich, 2004). The ensemble classifiers, which are built from a diverse feature subset randomly picked from the original feature set, accept the majority vote of the learners. Strong classifiers, which are frequently employed in the diagnosis of various illnesses and the classification of biomedical signals, can be used with ensemble learning techniques. Recently, ensemble classifiers have increasingly gained more attention in CAD applications (Bertoni et al., 2005; Das & Sengur, 2010; Kuncheva et al., 2010; Lai et al., 2006; Skurichina & Duin, 2002). Additionally, ensemble classifiers are naturally parallel, so if they have access to several processors during training and testing, they may perform better (Daumé III, 2012). Because of this, the unique contribution of this article is to create a trustworthy and accurate CAD scheme for categorizing EEG signals using ensemble classifier.

2.2 MATERIALS AND METHODOLOGY

2.2.1 DATASET DESCRIPTION

The CHB-MIT Scalp EEG Database (Shoeb, 2009; Goldberger et al., 2000) is a publicly available dataset of EEG by the Children's Hospital Boston and the Massachusetts Institute of Technology. The dataset was created to support research in epilepsy and seizure detection, and it contains EEG recordings from 22 patients ranging in age from 1 day to 58 years old who have experienced seizures. The recordings were performed during routine clinical care, and patients were not subjected to any additional testing or procedures. The dataset consists of continuous EEG recordings of up to 23 channels for each patient, sampled at 256 Hz for a duration of up to 23 hours. The recordings were performed using a variety of electrodes, including subdermal needle electrodes, scalp electrodes, and sintered Ag/AgCl electrodes. The dataset also includes annotations of the recordings, which indicate the presence of seizures and other events.

Database signals are separated into ictal and interictal folders. Every seizure has a preictal period of about one hour, so we combined preictal and ictal files and created 48 minutes of preictal EEG signal. For interictal signals, a file represents 1 hour of EEG signal, while for preictal signals, it depends on the length of the seizure. All patients have 24 or 25 hours, and we used 15 hours for the training dataset and the remaining 9 or 10 hours for the test phase. For the training data, we generally used a randomly chosen chunk of 10 minutes, but for one patient we had to use all interictal signals for the training dataset. We extracted 1000 interictal, 1000 ictal, and 1000 preictal seizure segments in total, extracting 8 second preictal segments for the period between 30 and 15 minutes prior to each seizure onset.

2.2.2 PROPOSED METHODOLOGY

The overall diagram of the proposed diagnosis methodology is presented in Figure 2.1. The methodology involves three fundamental stages: i) eliminating noise from signals; ii) extracting features from noiseless signals; iii) reducing the dimension of the

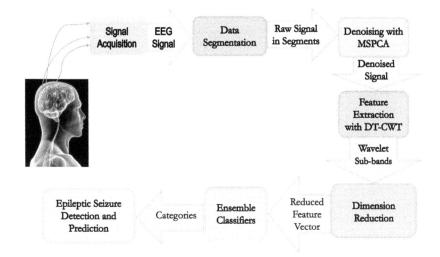

FIGURE 2.1   The overall diagram of the proposed methodology.

extracted feature set through feature selection; and iv) diagnosing epileptic seizure with ensemble classification models. The stages of the proposed methodology are briefly described as follows.

2.2.2.1   Noise Elimination of Signals with Multiscale Principal Component Analysis

Noise elimination of signals is an essential task in various applications ranging from signal processing to data analysis. Multiscale principal component analysis (MSPCA) is a powerful method for removing noise from signals. It is a combination of wavelet analysis and principal component analysis (PCA) that takes advantage of the strengths of both methods. Wavelet analysis decomposes a signal into different frequency components, known as scales; each scale contains information about the signal at a specific frequency range. The wavelet transform virtually de-correlates the relationship between the stochastic measurements of a signal. On the other hand, PCA identifies the most significant sources of variation in a dataset and removes the correlations between the variables.

By integrating the outcomes at relevant scales, MSPCA determines the PCA of the wavelet coefficients at each scale. This process removes relationships between variables by extracting linear relationships. MSPCA identifies the coefficients with the highest relevance by thresholding the input and output latent variables and selects the components with the same number at each scale so that the wavelet decomposition won't have an impact on the relationship between the variables. MSPCA can effectively remove noise from signals even when the noise is non-white or scale-dependent by using different PCA models for the coefficients at each scale and by applying thresholding criteria based on the nature of the application. MSPCA is particularly useful for processing signals that incorporate contributions from events whose behaviors change over time and in frequency (Malinowski, 2002).

MSPCA has been successfully applied to various real-world problems and has been shown to produce more accurate parameters, higher prediction abilities, and models that offer more physical insight into the system being modeled. Despite its effectiveness, a robust statistical foundation for the modeling and thresholding procedures is still needed to solve the optimization problem that arises in combining these two processes (Alsberg et al., 1998; Bakshi, 1998; Trygg & Wold, 1998; Bakshi et al., 1997).

2.2.2.2 Feature Extraction with DT-CWT

The wavelet transform is used to break down the EMG signals into a collection of fundamental operations known as wavelets, which improves time and frequency resolution. Shift variant functionality is a characteristic of the standard DWT due to the decimation process employed in the transform. Hence, a various set of wavelet coefficients present at the output can happen as the small change in the input signal is realized. Because of this, Kingsbury introduced the DT-CWT, a novel wavelet transform (Kingsbury, 1998), as a contemporary upgrade of the DWT with substantial new features.

In two dimensions, DT-CWT is directionally selective and nearly shift invariant. Even though the multidimensional DT-CWT is computationally efficient and separable in nature, it still depends on a filter bank (Daubechies, 1992). The DT-CWT makes use of two actual DWTs, one that provides the transform's real portion and another that provides the transform's imaginary portion. To achieve a transform that is practically analytical, two sets of filters are merged (Selesnick et al., 2005). The parameters are selected automatically with a seven-level decomposition in the DT-CWT implementation. As a consequence, each signal has 95 features completely extracted.

2.2.2.3 Dimension Reduction of Extracted Features

The dimension of the signal is reduced and statistical values of wavelet coefficients such as mean, standard deviation, skewness, and kurtosis are calculated using DT-CWT. Although first- and second-order statistics are crucial in signal processing, they are insufficient for many signals such as nonlinear signals like EEG. In order to better characterize such signals, higher-order statistics need to be considered. However, feature vectors produced by wavelet-based extraction methods are often too large to be used in classification tasks (Mendel, 1991).

Therefore, dimension reduction techniques are employed to extract fewer characteristics from wavelet coefficients. The reduced feature sets are derived from sub-bands of the wavelet decomposition using first-, second-, third-, and fourth-order cumulants of each level of sub-bands. A smaller feature set is desirable in classification tasks, as it provides a better characterization of the behavior of EEG signals. For EEG signal classification, six statistical features are used: mean absolute values, average power, standard deviation, ratio of absolute mean values of coefficients of adjacent sub-bands, skewness, and kurtosis of coefficients in each sub-band (Kevric & Subasi, 2017).

2.2.2.4 Ensemble Models

Ensemble modeling aims to create and merge multiple inductive models within a single domain to generate predictions that are superior to those from most or all individual models. By retaining or enhancing the strengths of specific models while reducing or eliminating their weaknesses, the overall performance can be improved. Surprisingly, even dozens or hundreds of models, some of which may be of subpar quality, can collaborate effectively to produce excellent forecasts. Both classification and regression, the two main predictive modeling tasks, can benefit from ensemble modeling. Although each individual model may be easily understood by humans, employing an ensemble approach may result in a significant boost in performance at the expense of greater computational resources for developing multiple models and a reduced level of overall interpretability.

Appropriate strategies for base model development and aggregation are needed in order to take advantage of this potential for improved predictive power. The majority of this material applies to both the classification and regression tasks because the former is primarily task independent, and the latter's task-specific elements are sufficiently simple and isolated. Ensemble models are, however, more frequently and effectively employed in the former case, and certain ensemble modeling strategies created especially for classification will also be explored (Cichosz, 2014). In this study we apply the following ensemble models:

The **bagging** (bootstrap aggregating) technique represents a straightforward ensemble modeling algorithm that integrates fundamental approaches to constructing and assembling base models: The construction of base models involves the utilization of bootstrap samples from the instruction set. The classification task can be accomplished through plain voting, while the regression task can be achieved through averaging. Class label voting and class probability averaging can result in a probabilistic form of bagging when utilizing probabilistic base classification approaches. This system, while not guaranteeing extreme prediction capabilities so long as the algorithm is unstable, is anticipated to provide an enhancement over standalone models constructed using the same methodology as base models. When base models are not sufficiently diversified for stable algorithms, there may be little improvement or even a slight decline in prediction quality. Other than instability, there are no specific constraints for the modeling algorithm. In fact, if doing so makes it more unstable, it might be simplified in comparison to what would typically be utilized to create a single model. Inappropriate overfitting safeguards may be abandoned, for example regression trees or pruning decision, in some algorithms. Models that are too closely matched to their specific bootstrap units are more likely to vary. The overfitting of base models will not necessitate the overfitting of the ensemble, as their aggregation will essentially cancel it out. Similar to attribute selection, there is typically no need to worry about it because more attributes offer more chances to develop a variety of models. In contrast to what is typically the case when single models are developed, this is a noticeable distinction. One way to think of bagging is as a way to make unstable algorithms more stable. Depending on a specific training set, individual models produced using such methods may be subject to significant variation. A better or worse model could potentially be produced for a slightly different training set. A useful answer cannot be achieved by building many models from various data samples without merging them into an ensemble and then picking the one that seems to work best. This is due to the fact that selecting a model depends on its evaluation, and model evaluation just makes sense when a repeatable modeling process is employed. In particular, it is necessary to repeat training and assessment cycles numerous times in order to provide low-variance performance estimates that could be used to select a model. For models that merely differ in their training units, this is categorically rejected. In an optimistic situation, bagging with a sufficient number of base models allows one to be quite confident that the final model is at least as effective as a single model, and most likely even better. If computing resources permit the generation of dozens or more models, as they typically do for this method, and if model human readability is not required, then bagging can be used. Up to a certain duration, the bagging ensemble performance tends to increase with the number of base models before stabilizing. The maximum amount of model diversity that can be achieved with bootstrap samples is attained at this point. In comparison to the other models, additional models are too numerous to make a difference (Cichosz, 2014).

AdaBoost stands for adaptive boosting and was first proposed by Freund and Schapire (1995). The algorithm works by iteratively building a set of weak classifiers on a training dataset; a weak classifier is a model that performs only slightly better than random guessing. The weak classifiers are combined to form a strong classifier that can make accurate predictions on new data. In each iteration, the algorithm assigns a weight to each sample in the training dataset. The weights are initially set to be uniform, but they are adjusted after each iteration to focus more on the samples that are misclassified by the previous weak classifier. The weight of each weak classifier is also adjusted based on its accuracy, with more weight given to more accurate classifiers. The final ensemble model is created by combining the weak classifiers based on their weights. When making a prediction, the weak classifiers' predictions are combined, with more weight given to the predictions of more accurate classifiers.

Multiboost is a boosting technique in which multiple weak learners are combined to create a strong learner. In Multiboost, decision stumps (simple decision trees with a single decision node) are used as weak learners. These decision stumps are trained on a subset of the data and their outputs are combined to create the final prediction. The algorithm works by iteratively training decision stumps on different subsets of the data, with each iteration focusing on the samples that were misclassified in the previous iteration. The weights of the misclassified samples are increased, so that the next decision stump will pay more attention to these samples. The final prediction is made by combining the outputs of all the decision stumps. Multiboost is a flexible algorithm that can be used for a wide range of classification tasks.

Random subspace (RS) is an ensemble learning technique used to improve the performance and robustness of machine learning models (Ho, 1998). It is based on the concept of generating multiple models using random subsets of the original features of the data. The random subspace method involves selecting a random subset of features (columns) from the training dataset to create a new dataset with fewer features. This is typically done by selecting a fixed number of features, or by specifying a percentage of the total features to be included in the subspace. Multiple models are then trained using different random subsets of the features, with each model being trained on a different subspace of the original features. Each model is trained independently using a base learning algorithm, such as decision trees, neural networks, or support vector machines. The final ensemble is created by combining the predictions of all the models, usually using a simple majority voting scheme for classification tasks or averaging for regression tasks. The combination of multiple models trained on different feature subsets helps to reduce the variance of the final prediction and improve the generalization performance of the model.

Rotation forest (RoF) is an ensemble learning method that combines random subspace method and PCA for feature selection. The method was introduced by Rodriguez et al. (2006). In 2006. In RoF, the original dataset is randomly partitioned into K disjoint subsets; PCA is then performed on each subset to reduce the dimensionality of the data. The principal components obtained from each subset are concatenated to form a new feature set. This process is repeated B times, resulting in B different feature sets. For each feature set, a decision tree classifier is built. The final ensemble is obtained by aggregating the predictions of all B trees.

The rationale behind RoF is that by using different subsets of features, the ensemble can capture different aspects of the data, leading to a more diverse and accurate model. Moreover, PCA is used to remove irrelevant and redundant features, reducing the risk of overfitting. RoF has been shown to be effective in various applications, including gene expression data analysis, text classification, and image recognition. However, it can be computationally expensive due to the multiple iterations of PCA and decision tree construction.

2.3 RESULTS AND DISCUSSION

2.3.1 EXPERIMENTAL DESIGN

For the experiment, we first checked the number of seizures for each patient. After that we divided signals into 2048 chunks and generated new ones with those parts. Specifically, we divided the channel of interictal and preictal data from the training dataset into smaller segments (2048 samples of 8 seconds each) and denoised the matrix using MSPCA. We extracted DT-CWT features from each denoised 8 second segment and put them into the training database. We then performed 10-fold cross validation on the training database to tune the machine learning algorithm parameters.

For each sub-band of DT-CWT, we retrieved statistical features (first-, second-, third-, and fourth-order moments) to detect epileptic seizures and predict them, using features from each frame of an EEG signal to describe the signal patterns. The magnitude of a feature is likely to differ significantly from one individual to another due to an EEG signal characteristic. It is important for a classification method to accommodate the variations that are expected. In this case, we first analyzed an EEG signal pattern to identify a set of features and then used single and ensemble classification algorithms to reduce errors in recognizing EEG signals.

Testing the performance of a classifier on a separate test set will be more reliable than testing it on the training set. It remains a contentious question when examining an algorithm's classification performance on a small set of data. We used 10-fold cross-validation in this study since it is an appropriate option for scenarios with little data. Typically, this method is used to represent each class in about the same proportions as in the entire dataset; it entails dividing the input data into 10 independent sections at random. In the end, the 10 error estimates are averaged to obtain an overall error estimate.

With the following four metrics, the confusion matrix plays a crucial role in assessing a classifier's performance: (1) true positives (TP), (2) true negatives (TN), (3) false negatives (FN), and (4) false positives (FP). For this study, we calculated performance verification criteria based on the defined four metrics: (1) accuracy, (2) F-measure, (3) kappa, and (4) AUC (area under the curve).

2.3.2 EXPERIMENTAL RESULTS

The results for the four metrics are presented in Tables 2.1 and 2.2, where the red and green colored cells respectively indicate maximum and minimum performance and the other colored cells represent a performance degree between them. Our proposed methodology obtained at least 97% accuracy in almost all cases: The best classification performance was achieved for all the ensemble models combined with the SVM classifier; accuracy was 95% only with Adaboost and Multiboost combined with random tree. This means the proposed methodology is well-designed and robust such that it can perform well independent of any ensemble classification model. Our methodology was also confirmed in that we achieved an AUC of 1, the maximum. Therefore, we concluded that our proposed methodology performs well in detecting epileptic seizures.

2.4 CONCLUSIONS

The goal of this study was to develop a methodology for analyzing EEG signals to detect and predict epileptic seizures. In the first stage of the introduced methodology, to reduce impulsive noise in raw EEG recordings, we applied MSPCA to remove noise from electrical and electronic equipment, movement artifacts, and other sources. Typically, EEG signals have a large number of dimensions, leading to difficulties in data analysis. Therefore, in the second stage of the approach, we applied DT-CWT to extract meaningful and instructive characteristics from EEG signals. We then described the distribution of wavelet coefficients by computing the statistical values of the DT-CWT sub-bands and input the statistical values of each DT-CWT sub-band into a classifier. It is worthwhile to note that ensemble classifiers have rarely been used in the literature for EEG data classification.

We verified the performance of the proposed methodology on a publicly available benchmark, the CHB-MIT Scalp EEG dataset, using a variety of ensemble learning algorithms combined with single classification. The results indicate that the methodology performs well for all ensemble classifiers. To be specific, when combined with SVM, it achieves an outstanding classification accuracy of 99.63%. We conclude that the proposed methodology showed satisfactory performance in detecting and predicting epileptic seizures due to the following reasons: 1) The methodology is a well-designed and well-built framework: 2) MSPCA is a suitable preprocessing method for denoising signals; and 3) DT-CWT can extract high-informative features that well represent the characteristics of signals.

2.5 FUNDING

This work was supported by Effat University (Decision Number UC#7/28 Feb. 2018/10.2–44i), Jeddah, Saudi Arabia.

TABLE 2.1
Accuracy and F-Measure of Ensemble Classifiers

	Accuracy					F-measure				
	Bagging	Adaboost	Multiboost	RS	RoF	Bagging	Adaboost	Multiboost	RS	RoF
SVM	99.63%	99.53%	99.53%	99.63%	99.53%	0.996	0.995	0.995	0.996	0.995
k-NN	97.73%	97.07%	97.20%	97.87%	98.27%	0.977	0.971	0.972	0.979	0.983
ANN	99.43%	99.37%	99.27%	99.40%	99.50%	0.994	0.994	0.993	0.994	0.995
Random Forest	98.90%	99.03%	99.03%	98.97%	99.20%	0.989	0.990	0.990	0.990	0.992
C4.5	98.20%	99.17%	98.83%	98.73%	99.60%	0.982	0.992	0.988	0.987	0.996
REP Tree	97.47%	99.17%	98.87%	98.47%	98.73%	0.975	0.992	0.989	0.985	0.987
Random Tree	98.27%	95.60%	95.60%	98.77%	98.93%	0.983	0.956	0.956	0.988	0.989
Simple Logistic	99.43%	99.43%	99.43%	99.23%	99.50%	0.994	0.994	0.994	0.992	0.995

TABLE 2.2

AUC and Kappa Statistics for Ensemble Classifiers

	AUC					Kappa				
	Bagging	Adaboost	Multiboost	RS	RoF	Bagging	Adaboost	Multiboost	RS	RoF
SVM	0.999	0.997	0.997	0.999	0.999	0.995	0.993	0.993	0.995	0.993
k-NN	0.996	0.983	0.990	0.998	0.998	0.966	0.956	0.958	0.968	0.974
ANN	1.000	0.998	0.995	1.000	1.000	0.992	0.991	0.989	0.991	0.993
Random Forest	1.000	1.000	1.000	1.000	1.000	0.984	0.986	0.986	0.985	0.988
C4.5	0.999	1.000	1.000	1.000	1.000	0.973	0.988	0.983	0.981	0.994
REP Tree	0.998	1.000	1.000	1.000	1.000	0.962	0.988	0.983	0.977	0.981
Random Tree	0.999	0.967	0.967	0.999	0.999	0.974	0.934	0.934	0.982	0.984
Simple logistic	1.000	0.999	0.998	1.000	1.000	0.992	0.992	0.992	0.989	0.993

REFERENCES

Adeli, H., Ghosh-Dastidar, S., & Dadmehr, N. (2007). A wavelet-chaos methodology for analysis of EEGs and EEG subbands to detect seizure and epilepsy. *IEEE Transactions on Biomedical Engineering*, *54*(2), Article 2.

Alam, S. S., & Bhuiyan, M. I. H. (2013). Detection of seizure and epilepsy using higher order statistics in the EMD domain. *IEEE Journal of Biomedical and Health Informatics*, *17*(2), 312–318.

Alsberg, B. K., Woodward, A. M., Winson, M. K., Rowland, J. J., & Kell, D. B. (1998). Variable selection in wavelet regression models. *Analytica Chimica Acta*, *368*(1), 29–44.

Bakshi, B. R. (1998). Multiscale PCA with application to multivariate statistical process monitoring. *AIChE Journal*, *44*(7), Article 7.

Bakshi, B. R., Bansal, P., & Nounou, M. N. (1997). Multiscale rectification of random errors without fundamental process models. *Computers & Chemical Engineering*, *21*, S1167–S1172.

Bertoni, A., Folgieri, R., & Valentini, G. (2005). Feature selection combined with random subspace ensemble for gene expression based diagnosis of malignancies. *Biological and Artificial Intelligence Environments*, 29–35.

Birjandtalab, J., Pouyan, M. B., Cogan, D., Nourani, M., & Harvey, J. (2017). Automated seizure detection using limited-channel EEG and non-linear dimension reduction. *Computers in Biology and Medicine*, *82*, 49–58.

Blum, D., Eskola, J., Bortz, J., & Fisher, R. (1996). Patient awareness of seizures. *Neurology*, *47*(1), Article 1.

Brodie, M., Barry, S., Bamagous, G., Norrie, J., & Kwan, P. (2012). Patterns of treatment response in newly diagnosed epilepsy. *Neurology*, doi: 10.1212/WNL-0b013e3182563b19.

Carrette, S., Boon, P., Sprengers, M., Raedt, R., & Vonck, K. (2015). Responsive neurostimulation in epilepsy. *Expert Review of Neurotherapeutics*, *15*(12), 1445–1454.

Cichosz, P. (2014). *Data mining algorithms: Explained using R*. John Wiley & Sons.

Das, R., & Sengur, A. (2010). Evaluation of ensemble methods for diagnosing of valvular heart disease. *Expert Systems with Applications*, *37*(7), 5110–5115.

Daubechies, I. (1992). *Ten lectures on wavelets*. SIAM.

Daumé III, H. (2012). A course in machine learning. *Chapter*, *5*, 69.

Fisher, R. S., Krauss, G. L., Ramsay, E., Laxer, K., & Gates, J. (1997). Assessment of vagus nerve stimulation for epilepsy: Report of the therapeutics and technology assessment subcommittee of the American Academy of Neurology. *Neurology*, *49*(1), Article 1.

French, J. A., & Pedley, T. A. (2008). Initial management of epilepsy. *New England Journal of Medicine*, *359*(2), Article 2.

Freund, Y., & Schapire, R. E. (1997). A desicion-theoretic generalization of on-line learning and an application to boosting. *Journal of Computer and System Sciences*, *55*(1), 119–139.

Goldberger, A. L., Amaral, L. A., Glass, L., Hausdorff, J. M., Ivanov, P. C., Mark, R. G., Mietus, J. E., Moody, G. B., Peng, C.-K., & Stanley, H. E. (2000). Physiobank, physiotoolkit, and physionet. *Circulation*, *101*(23), Article 23.

Ho, T. K. (1998). The random subspace method for constructing decision forests. *IEEE Transactions on Pattern Analysis and Machine Intelligence*, *20*(8), Article 8.

Hosseini, M.-P., Pompili, D., Elisevich, K., & Soltanian-Zadeh, H. (2018). Random ensemble learning for EEG classification. *Artificial Intelligence in Medicine*, *84*, 146–158.

Iasemidis, L. D. (2003). Epileptic seizure prediction and control. *IEEE Transactions on Biomedical Engineering*, *50*(5), Article 5.

Jayapandian, C., Chen, C.-H., Dabir, A., Lhatoo, S., Zhang, G.-Q., & Sahoo, S. S. (October 2014). Domain ontology as conceptual model for big data management: Application in biomedical informatics. *International Conference on Conceptual Modeling*, Atlanta, USA, 144–157.

Kevric, J., & Subasi, A. (2017). Comparison of signal decomposition methods in classification of EEG signals for motor-imagery BCI system. *Biomedical Signal Processing and Control*, *31*, 398–406.

Kingsbury, N. G. (September 1998). The dual-tree complex wavelet transform: A new technique for shift invariance and directional filters. *Proceedings of the European Signal Processing Conference Rhodes*, *8*, 86, 319–322.

Kuncheva, L. I. (2004). *Combining pattern classifiers: Methods and algorithms*. John Wiley & Sons.

Kuncheva, L. I., Rodríguez, J. J., Plumpton, C. O., Linden, D. E., & Johnston, S. J. (2010). Random subspace ensembles for fMRI classification. *IEEE Transactions on Medical Imaging*, *29*(2), 531–542.

Lai, C., Reinders, M. J., & Wessels, L. (2006). Random subspace method for multivariate feature selection. *Pattern Recognition Letters*, *27*(10), 1067–1076.

Li, X., & Yao, X. (September 2005). Application of fuzzy similarity to prediction of epileptic seizures using EEG signals. *International Conference on Fuzzy Systems and Knowledge Discovery*, China, 645–652.

Litt, B., & Echauz, J. (2002). Prediction of epileptic seizures. *The Lancet Neurology*, *1*(1), 22–30.

Malinowski, E. R. (2002). *Factor analysis in chemistry*. Wiley.

Mendel, J. M. (1991). Tutorial on higher-order statistics (spectra) in signal processing and system theory: Theoretical results and some applications. *Proceedings of the IEEE*, *79*(3), 278–305.

Pack, A. M. (2012). SUDEP: What are the risk factors? Do seizures or antiepileptic drugs contribute to an increased risk? *Epilepsy Currents*, *12*(4), Article 4.

Rajaei, H., Cabrerizo, M., Janwattanapong, P., Pinzon-Ardila, A., Gonzalez-Arias, S., & Adjouadi, M. (August 2016). Connectivity maps of different types of epileptogenic patterns. *Annual International Conference of the IEEE Engineering in Medicine and Biology Society*, 1018–1021.

Rodriguez, J. J., Kuncheva, L. I., & Alonso, C. J. (2006). Rotation forest: A new classifier ensemble method. *IEEE Transactions on Pattern Analysis and Machine Intelligence*, *28*(10), 1619–1630.

Saeedi, R., Fallahzadeh, R., Alinia, P., & Ghasemzadeh, H. (August 2016). *An energy-efficient computational model for uncertainty management in dynamically changing networked wearables*. *Proceedings of the 2016 International Symposium on Low Power Electronics and Design*, New York, NY, 46–51.

Selesnick, I. W., Baraniuk, R. G., & Kingsbury, N. C. (2005). The dual-tree complex wavelet transform. *IEEE Signal Processing Magazine*, *22*(6), 123–151.

Shoeb, A. H. (June 2009). Application of machine learning to epileptic seizure detection. *ICML'10: Proceedings of the 27th International Conference on International Conference on Machine Learning*, Hafia, Israel, 975–982.

Skurichina, M., & Duin, R. P. (2002). Bagging, boosting and the random subspace method for linear classifiers. *Pattern Analysis & Applications*, *5*(2), 121–135.

Trygg, J., & Wold, S. (1998). PLS regression on wavelet compressed NIR spectra. *Chemometrics and Intelligent Laboratory Systems*, *42*(1–2), 209–220. https://doi.org/10.1016/S0169-7439(98)00013-6

Valentini, G., & Dietterich, T. G. (2004, Jul). Bias-variance analysis of support vector machines for the development of SVM-based ensemble methods. *Journal of Machine Learning Research*, *5*, 725–775.

3 Classification of Normal and Alcoholic EEG Signals Using Signal Processing and Machine Learning

Fatima Faraz, Mohammad Ebad Ur Rehman, Gary Tse, and Haipeng Liu

3.1 DIAGNOSIS OF ALCOHOLISM: AN UNMET CLINICAL NEED

One of the leading causes of preventable noncommunicable diseases is material abuse, and alcohol constitutes a significant proportion (1). According to the latest comparative risk assessment done by the World Health Organization (WHO), alcohol consumption is the third leading global risk for the burden of disease as measured in disability-adjusted life years (2). The harmful use of alcohol is a causal factor in more than 200 disease and injury conditions, leading to 3 million deaths every year globally.

Alcohol use disorder (AUD) refers to a condition where a person experiences a lack of control over their alcohol consumption, resulting in physical dependence and tolerance. This in turn leads to harmful effects on their psychological, social, and physical well-being (3, 4). Not only can AUD cause liver and pancreatic diseases but also various types of cancer and mental illnesses (5, 6). AUD is a leading co-occurring disorder in people with severe mental illnesses like schizophrenia and bipolar disorder (3, 7). There are lower risks of cardiovascular and all-cause mortality in people with low levels of alcohol consumption when compared with abstainers, whereas the cardio-protective role significantly diminishes in heavy drinkers (6). The accurate and early detection of alcoholism plays a key role in effective intervention and management of AUD-associated disorders and diseases.

The early diagnosis of alcoholism can be challenging due to patients' embarrassment and denial, the presence of nonspecific symptoms, and misconceptions about the condition (8). The fifth edition of the *Diagnostic and Statistical Manual of Mental Disorders* defined AUD as a cluster of behavioral and physical symptoms with a continuum or spectrum of severity (5). There is a lack of specific diagnostic criteria for AUD. In clinical practice, a physician needs to go beyond laboratory

DOI: 10.1201/9781003479970-3

results and clinical findings, gather socioeconomic and psychiatric information, and comprehensively analyze the effects of different factors on a patient's life.

3.2   ALCOHOLISM-RELEVANT CHANGES IN EEG WAVES

Electroencephalography (EEG) involves the detection of brain activities using small electrodes attached to the scalp. Compared to other neuroimaging techniques such as magnetoencephalography, functional magnetic resonance imaging (fMRI), functional near-infrared spectroscopy (fNIRS) and positron emission tomography, EEG is often preferred since it is noninvasive, nonradioactive, low-cost, easy to perform, and sensitive to various neural activities. EEG has been widely applied in neuroscience and the diagnosis of some neurological diseases (9). Evidence showed that alcoholism can generate changes in various EEG features, e.g., power spectral intensity, coherence, event-related potentials, and sensory pathway potentials (10, 11), which in turn provides the possibility of EEG-based detection of alcoholism (12, 13).

Theta waves are frequently found in awake adult EEGs, particularly in the frontocentral regions. Consistently elevated theta power in a single region is indicative of underlying functional and structural changes, which has been applied in the diagnosis of psychiatric disorders (14). Rangaswamy et al. undertook a case control study to compare theta power between alcoholics and age- and gender-matched controls. Theta power was significantly increased in male alcoholics at the central and parietal loci and in female alcoholics at the parietal loci (15). The elevated theta power may be indicative of a decrement in information processing capacity of the brain, or it may highlight the imbalance between cortical excitation and inhibition (10, 16, 17).

The alpha wave is the principal rhythm in relaxed awake adult EEG patterns. A decline in the relative power of alpha waves is observed in alcoholics compared with healthy controls, particularly in the occipital foci. This decline is associated with a reduction in attention and ability to retrieve information from learned memory (16). In addition, low-voltage alpha waves were more common in alcoholics than in healthy controls ($< 10\ \mu V$) (18).

Several studies have demonstrated that an increase in beta power, particularly in fast beta power (> 20 Hz), is associated with both alcoholism and family history of alcoholism (19, 20). The elevation in beta power was more prominent in male alcoholics (21). De Bruin et al. demonstrated impaired synchronization in both males and females at the alpha and beta frequencies, with the impairment increasing with increasing intake (22). Drinking variables such as recency and amount of alcohol consumption did not alter the observed patterns (10, 16, 21). Meanwhile, several confounders influence the discriminatory power of beta waves. Medication, particularly sedative-hypnotics, epilepsy, hallucinations, and family history of alcoholism have been reported to significantly enhance beta power (10).

The absolute gamma band power is altered in alcoholism (23). Individuals who are susceptible to AUD, regardless of gender and ancestry, demonstrate lower gamma activity in posterior regions (such as the occipital and parietal lobes) (24).

Brain rhythms were taken into consideration in detecting alcoholism on EEG, and the gamma band (30–40 Hz) was found to be the most significant rhythm (25).

EEG patterns may serve to predict relapse in patients undergoing detoxification. Winterer et al. observed more desynchronized beta activity over frontal areas in relapsing alcoholics compared with non-relapsers, which suggests a functional disturbance in the prefrontal cortex (26). Bauer enrolled and followed 107 substance-dependent individuals from residential treatment programs and found that beta power was significantly greater in patients who relapsed than in successfully abstaining individuals (27). In a similar study comparing relapse-prone patients, abstaining patients, and healthy controls, Saletu-Zyhlarz et al. observed a significant elevation in the theta and delta power in the patients who eventually relapsed (16). The EEG patterns were comparable in patients who abstained successfully for the six-month follow-up period and the healthy controls. Alpha waves were reported to have the opposite pattern, with an increase in spectral power being consistently observed in abstainers but not relapse-prone individuals (16).

3.3 POTENTIAL OF AI-ASSISTED EEG ANALYSIS FOR AUTOMATIC DIAGNOSIS

Visual analysis of EEG by experts is still deemed the gold standard since it outperforms current computer algorithms (28, 29). However, visual analysis is time-consuming. Moreover, highly trained staff are required; it also remains error prone, limiting its applicability for discriminating alcoholics from non-alcoholics. Quantitative EEG (qEEG) analysis has been proposed to overcome the inherent limitations of visual EEG analysis, which include the inter-rater variability along with the low sensitivity in recognizing subtle abnormalities, providing the possibility of automatic diagnosis of neurological and psychological disorders including alcoholism (30).

Analyzing qEEG depends on selecting and extracting relevant features, which needs proper algorithms. To achieve clinically applicable diagnostic tools, it is essential to develop automatic algorithms for qEEG analysis that can handle large-scale EEG data. Recently, there is an increasing interesting in artificial intelligence (AI)-assisted EEG analysis. Based on large datasets, AI methods facilitate accurate and reliable interpretation of EEG data for robust decision making (31). As shown in Figure 3.1, the steps of AI-assisted EEG analysis include data acquisition, pre-processing, feature extraction, training of the algorithm, classification, and evaluation of performance. The feature extraction and classification play key roles in the whole process. For example, through time-frequency analysis, changes in the frequency of EEG signals and in spatiotemporal features can be acquired over time to describe activities of various parts of the brain. The changes in EEG feature that reflect alcoholism-related neural activities can be input in AI algorithms for the classification between alcoholic and normal subjects (32). To provide an overview of the state of the art and an in-depth understanding of the research gaps, in the following sections, the main machine learning and deep learning algorithms will be illustrated.

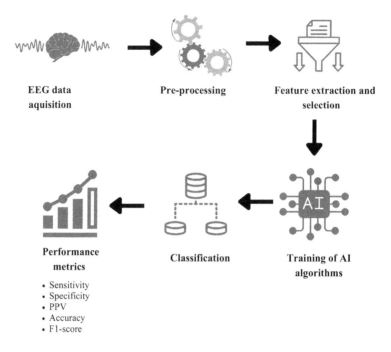

FIGURE 3.1   Flowchart of AI-assisted automatic EEG analysis.

3.4   MACHINE LEARNING IN EEG-BASED AUTOMATIC DETECTION OF ALCOHOLISM

3.4.1   DATASET

The EEG database of the University of California, Irvine Knowledge Discovery in Database (UCI KDD) Archive is used in the majority of recent studies (https://kdd. ics.uci.edu/databases/eeg/eeg.data.html) (Table 3.1). This dataset was initially developed to examine genetic predisposition, through EEG signals, to alcoholism (33). The dataset has three available versions: small, large, and full. The small dataset contains data of two subjects, one alcoholic and one normal. The large dataset contains training and testing data for 10 alcoholic and 10 control subjects. The full dataset contains all 120 trials for 122 subjects (77 alcoholic and 45 normal). The normal subjects in the control group had no particular or family history of alcohol misuse or neurological disorder or any history of psychiatric disease. The full UCI KDD is the mostly commonly used in recent machine learning methods (Table 3.1). For instance, Bae et al. screened 60 subjects (37 alcoholics, 23 normal) were screened from the full UCI KDD (34), and others have used the small (35) and large (36) sets.

Some researchers introduced new EEG datasets. To achieve fine-grained classification of alcohol abusers, alcoholics, and normal subjects, Mumtaz et al. collected EEG and other clinical data from 45 subjects (12 alcohol abusers, 18 alcoholics, and 15 healthy controls) in University Malaya Medical Center (20). To identity biomarkers indicating a predisposition to AUD, Kinreich et al. used the Collaborative Study of the Genetics of Alcoholism (COGA) dataset and selected 656 subjects who were

TABLE 3.1

Machine Learning Methods for the Automated EEG Diagnosis of Alcoholism

Study	Dataset	EEG features	Classification algorithm	Performance
Acharya et al., 2012 (46)	UCI KDD, 122 subjects (77 alcoholic and 45 normal)	approximate entropy, largest Lyapunov exponent, sample entropy, four higher-order spectra features	SVM	accuracy: 91.7%, sensitivity: 90%, specificity: 93.3%,
Bae et al., 2017 (34)	UCI KDD (screened), 60 subjects (37 alcoholics, 23 normal)	clustering coefficient, assortativity, average neighborhood degree, node betweenness centrality	SVM	accuracy: 90%, sensitivity: 95.3%, specificity: 82.4%
Zhu et al., 2014 (47)	UCI KDD, 122 subjects (77 alcoholic and 45 normal)	horizontal visibility graph entropy, sample entropy	SVM, KNN	accuracy: 95.8 %, using selected 13 features by KNN
Kinreich et al., 2021 (24)	Collaborative Study on the Genetics of Alcoholism Dataset, 656 participants (328 alcoholic and 328 normal)	EEG features: spectral power (40 features), coherence (90 features), correlation (90 features) other features: single nucleotide polymorphisms (SNPs, 149 features), family history (2 features)	SVM	highest accuracy: 87.55% (specificity: 85.71%, sensitivity: 89.38%, AUC: 0.99, F1-Score: 0.89), based on EEG+SNP features
Zhang et al., 2020 (33)	UCI KDD 122 subjects (77 alcoholic and 45 normal)	traditional features: gray-level co-occurrence matrix transfer learning, local binary patterns, central moments features of two-dimensional EEG signals CNN combined with transfer learning	Gaussian Naive Bayes, KNN, MLP, RF, SVM	accuracy: 95.33%, precision: 95.68%, F1-Score: 95.24%, recall: 95.00%, achieved by SVM

(Continued)

TABLE 3.1 (Continued)
Machine Learning Methods for the Automated EEG Diagnosis of Alcoholism

Study	Dataset	EEG features	Classification algorithm	Performance
Buriro et al., 2021 (36)	UCI KDD: 20 alcoholics and 20 controls	features extracted by CNN and wavelet scattering transform	SVM-RBF, LDA	accuracy: 100.0%, sensitivity: 100.0%, specificity: 100.0%, AUC: 100.0%, achieved by SVM-RBF
Salankar et al., 2022 (45)	UCI KDD, 122 subjects (77 alcoholic and 45 normal)	second order difference plots of oscillatory modes and IMFs	LS-SVM, MLPNN, KNN, RF	accuracy: 99.89%, sensitivity: 99.56%, specificity: 99.67%, F1-score: 99.56%, by MLPNN
Anuragi et al., 2020 (35)	UCI KDD: one alcoholic and one control	70 features: mean, standard deviation, variance, skewness, kurtosis, Shannon entropy, and log entropy, from 10 IMFs	LS-SVM, SVM, KNN, naïve Bayes	average accuracy: 98.75%, sensitivity: 98.35%, specificity: 99.16%, precision: 99.17%, F-measure: 98.76%, MCC: 97.50%, by LS-SVM
Rodrigues et al., 2019 (40)	UCI KDD, 122 subjects (77 alcoholic and 45 normal)	For each sub-band: minimum, maximum, power, mean, standard deviation, and the absolute mean For adjacent sub-bands: the ratio of the absolute mean	SVM, KNN, MLP, OPF, naïve Bayes	99.87% for accuracy specificity, sensitivity, and positive predictive value, by naïve Bayes
Mumtaz et al. 2016 (20)	Collected from the University of Malaya Medical Center: 45 subjects (12 alcohol abusers, 18 alcoholics, and 15 healthy controls)	absolute power, relative power, sample entropy, approximate entropy	SVM, LDA, MLP, LMT	accuracy: 96%, sensitivity: 97%, specificity: 93%, by LMT

AUC: area under the curve, MCC: Matthews correlation coefficient, UCI KDD: University of California, Irvine Knowledge Discovery in Database, SVM: support vector machine, KNN: K-nearest neighbor, MLP: multilayer perceptron, RF: random forest, CNN: convolutional neural networks, RBF: radial basis function, LDA: linear discriminant analysis, LS-SVM: least square support vector machine, MLPNN: multilayer perceptron neural network, IMF: intrinsic mode functions, OPF: optimum-path forest, LMT: logistic model trees recruited as early as age 12 and were unaffected at first assessment and reassessed years later as having AUD (n = 328) or unaffected (n = 328) (24). This COGA dataset include rich details on EEG, genetic, and family history information, enabling the development of automatic alcoholism detection using multimodal data.

Various EEG datasets are available for brain–computer interface or sleep analysis that give researchers significant choice in terms of number, age, and gender of subjects, duration of the recordings, etc. In the case of alcoholism, however, UCI KDD is the only mainstream publicly available dataset (37). There is a need for more open-access EEG datasets to develop reliable machine learning methods of alcoholism detection that are applicable in different cohorts.

3.4.2 PREPROCESSING OF EEG SIGNALS

Analyzing EEG signals is challenging due to their low signal-to-noise ratios and nonstationary nature (32). Artifacts can occur in EEG recordings due to head movements, eye blinking, or electromyographic signals of neighboring muscles. The optimization of signal-to-noise ratio by preprocessing and feature extraction is a crucial aspect of automated EEG analysis (Figure 3.2). Narrow bandpass filters (e.g., Butterworth filters) are the traditional tool for de-noising EEG signals and separating different EEG rhythms (38). EEG signals are variable in both time and frequency, and time–frequency analysis enables both preprocessing and feature extraction in different domains; therefore, it has been adopted as a mainstream methodology (35, 39). The details will be summarized in the following subsection.

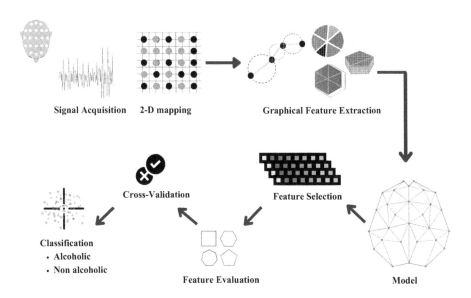

FIGURE 3.2 Framework of developing machine learning models for the EEG-based automatic detection of alcoholism.

3.4.3 EEG FEATURE EXTRACTION

As aforementioned, due to the nonlinear nature of EEG signals, scholars have developed time–frequency signal processing techniques for preprocessing and feature detection, such as fast Fourier transform (FFT), short-time Fourier transform, wavelet transform, Wigner–Ville distribution, Hilbert–Huang transform time-frequency distributions, eigenvector methods, and autoregression (35). Among these methods, wavelet analysis is the commonest. The wavelet transforms decompose EEG signals into discrete frequency sub-bands. Different wavelet families can be used in the decompositions of EEG signals, e.g., Biorthogonal, Coiflet, Daubechies, and Symlets (40).

Some advanced methods have been developed based on FFT and wavelet transform to suit EEG data and improve the efficiency of feature detection. For example, empirical wavelet transform uses an empirical method instead of dilation to decompose EEG recordings into certain sub-bands using adaptive bandpass filtering (35). Another example is Fourier–Bessel series expansion based empirical wavelet transform, which can decompose EEG signals into narrow sub-band signals using a boundary detection approach (41). To reveal the fluctuating frequencies of non-stationary EEG signals, Sadiq et al. used fractional Fourier transform with different coefficients (42). Buriro et al. used the wavelet scattering transform (WST), a recently developed knowledge-based feature extraction technique that is structurally like a convolutional neural network (CNN) (36). The WST preserves information in high frequency, is insensitive to signal deformations, and generates low-variance features of real-valued signals generally required in classification tasks (36).

Various features can be extracted from time–frequency analysis. To determine the optimal channels for EEG-based detection of alcoholism, Bavkar et al. used empirical mode decomposition (EMD) to extract the amplitude and frequency modulated bandwidth features from the intrinsic mode function (IMF) (43). Agarwal and Zubair used sliding singular spectrum analysis to decompose and de-noise EEG signals, using independent component analysis (ICA) to select the prominent alcoholic and non-alcoholic components from the preprocessed EEG data (44); the sliding SSA-ICA algorithm helps in reducing the computational time and complexity of the machine learning models. Salankar et al. used two robust methods, i.e., EMD and variational mode decomposition, to decompose EEG signals, and used second order difference plots to design features space (45).

3.4.4 ADVANCED EEG FEATURES OF ALCOHOLISM

Recent studies identified some EEG-based biomarkers of alcoholics, e.g., linear, nonlinear, and statistical features of alpha and gamma rhythms (38), as well as autocorrelation coefficients (39). Different entropies have been widely explored (46). For example, Zhu et al. applied horizontal visibility graph entropy and sample entropy on the full UCI KDD data and found that the former performed much better, providing a promising approach for alcoholism identification (13, 47).

In addition, coherence scores may be related to the heritable nature of alcoholism. Higher interhemispheric EEG coherence has been reported in the frontal and parietal foci of individuals with a family history of AUD than in controls with no apparent family history (48). Tcheslavski and Gonen reported a significant decrease in the coherence and phase synchronization in alcoholics as compared with controls (49). The correlation dimension acts as a good diagnostic parameter for the neural changes that occur in various neurological disorders including Parkinson's disease, Alzheimer's disease, glaucoma, schizophrenia, and alcoholism (50).

Alcoholism can influence the connectivity of different brain regions and their dynamics. New EEG features are proposed to reflect the topological changes. Wang et al. proposed a state space model that analyzes multichannel EEG signals and learns the dynamics of different sources corresponding to brain cortical activity; applying this model on full UCI KDD data, the authors found reveal significant attenuation of brain activity in response to visual stimuli in alcoholic subjects compared to healthy controls (51). Saqid et al. explored phase space dynamic and geometrical features for visualizing the chaotic nature and complexity of EEG signals and discovered 34 graphical features for decoding the chaotic behavior of normal and alcoholic EEG signals (52). Shen et al. used continuous wavelet transform and cross-mutual information (CMI) algorithms to estimate the functional connectivity maps between the whole brain regions. After the evaluation of all CMI connectivity values, the adjacent connectivities between the left parietal, left frontal, right temporal, right frontal, and right parietal were found to be the fuzzy locations in determining alcoholism (25).

Functional connectivity (FC) describes the statistical dependency between two regions, and effective connectivity (EC) estimates the cause–effect relationships between the activation patterns of different regions. Pain et al. used partial correlation, phase lag index, and Granger causality (GC) to estimate time FC, phase FC, and time EC to describe the topological functional-relationship patterns between brain regions in alcoholic and normal subjects, where the Beta band exhibited the highest classification accuracy (53). Similarly, Bae et al. calculated the GC between different EEG channels to construct effective networks. They extracted 10 graph-based parameters from the GC-derived networks that showed efficiency in differentiating alcoholic and normal subjects (34).

3.4.5 FEATURE SELECTION

The number of input variables in a machine learning algorithm is limited by the data size. To reduce the dimension of features and the computational complexity of the classification model, insignificant features need to be excluded. Conventional statistical tests, e.g., t-test (42, 46) and Mann–Whitney U test (45, 54), have been widely applied in selecting alcoholism-related EEG features. The features with p values less than 0.5 are selected for algorithm development while other features are excluded.

Meanwhile, some advanced feature selection methods have been proposed to identify alcoholism-related EEG features and channels. Rajaguru et al. used Hilbert transform, rigid regression, and chi-square probability density function to reduce the dimensionality of alcoholic EEG features (55). Kumari et al. used correlation-based feature selection and relief attribute rank selection to select five EEG features that afforded the accurate detection of alcoholism (56). Fayyaz et al. integrated the feature selection in preprocessing using peak visualization, which selects the peaks with distinctive width and height range and used machine learning techniques to find the most discriminating ranges; they then used the selected range peaks to compute features like indices of peaks, prominence of peak, contour heights, relative maxima, relative minima, local maxima, and local minima (57).

3.4.6 CLASSIFICATION: ALGORITHMS AND PERFORMANCE

The main traditional machine learning algorithms, e.g., support vector machine (SVM), K-nearest neighbor (KNN), multilayer perceptron, random forest (RF), and naive Bayes, as well as their variations, have been widely applied in EEG-based alcoholism detection (Table 3.1). Recent researchers have comprehensively compared different machine learning algorithms (**20, 33, 35, 40, 45**), providing valuable reference for future research. Some modified algorithms showed promising performance. For example, the least square support vector machine outperforms traditional methods like SVM and KNN (35, 56). The majority of recent studies show evaluation metrics above 0.90 (Table 3.1), indicating the applicability of AI-assisted EEG analysis to the automatic detection of alcoholism.

Some advanced classification algorithms appeared in recent studies. Bavkar et al. used an improved binary gravitational search algorithm (IBGSA) as an optimization tool to select the optimum EEG channels for the rapid screening of alcoholism. They used absolute gamma band power and KNN as a feature and ensemble subspace and a classifier, respectively. Providing a detection accuracy of 92.50% with only 13 EEG channels, IBGSA outperformed genetic algorithm, binary particle swarm optimization, and binary gravitational search algorithm (23). Dong and Wang proposed an algorithm for the classification of multidimensional datasets based on conjugate Bayesian multiple-kernel grouping learning (BMKGL). The proposed algorithms improved computational efficiency and enabled the integration of information from different dimensions to find the dimensions that most contributed to the variations in the outcome. Meanwhile, BMKGL can select the most suitable combination of kernels for different dimensions to extract the most appropriate measure for each dimension and improve the accuracy of classification results. BMKGL outperformed previous machine learning methods, i.e., KNN, SVM, and naive Bayes, in detecting alcoholism on the full UCI KDD EEG dataset (58).

3.5 DEEP LEARNING IN THE EEG-BASED AUTOMATIC DETECTION OF ALCOHOLISM

3.5.1 DATASETS AND PREPROCESSING

All the deep learning models shown in Table 3.2 were developed on the UCI KDD data. Of note, the large UCI KDD set (only 20 subjects) was used in some studies that involved multiple trials and segmentation of EEG data to generate a proper data size (25, 59, 60). Furthermore, Rahman et al. included only 10 subjects (5 alcoholic, 5 normal) who were randomly selected from the large UCI KDD set (61).

TABLE 3.2

Deep Learning Methods for Automated EEG Diagnosis of Alcoholism

Study	Dataset	EEG preprocessing	Classification algorithm	Performance
Rahman et al., 2020 (61)	UCI KDD, 10 subjects (5 alcoholic and 5 normal)	DWT, PCA, ICA	KNN, LVQ, RNN, B-LSTM, CNN	accuracy: 0.950, sensitivity: 0.945, specificity: 0.955, AUC: 0.950. by B_LSTM
Shen et al., 2022 (25)	UCI KDD, 20 subjects (10 alcoholic and 10 normal)	cross-mutual information based on the continuous wavelet transform	2D and 3D CNNs	accuracy: 96.25 ± 3.11% by 3D CNN
Fayyaz et al., 2019 (57)	UCI KDD, 122 subjects (77 alcoholic and 45 normal)	peak visualization method: selecting peaks with distinctive width and height range in order to compute features	LSTM	average accuracy of 5-fold cross-validation: 90%
Farsi et al., 2020 (62)	UCI KDD, 122 subjects (77 alcoholic and 45 normal)	PCA	1. ANN using PCA-derived features 2. LSTM using raw EEG data	average accuracy of 10-fold cross-validation: 86% for PCA-ANN, 90% for LSTM
Mukhtar et al., 2021 (37)	UCI KDD, 122 subjects (77 alcoholic and 45 normal)	normalization: scales the data so that it has a mean of 0 and a standard deviation of 1.	CNN	average accuracy: 98% on a batch size of 16 (range: 2^2 to 2^8)
Neeraj et al., 2021 (59)	UCI KDD, 20 subjects (10 alcoholic and 10 normal)	Preprocessing and feature extraction using moving window and FFT	a hybrid CNN-LSTM-ATTN network	accuracy: 98.83%
Kumari et al., 2022 (60)	UCI KDD, 20 subjects (10 alcoholic and 10 normal)	no preprocessing or feature extraction	CNN, LSTM and CNN + LSTM models	average classification accuracy on the testing dataset: 92.77%, 89%, and 91% along with error rates of 7.5%, 11.90%, and 8.7% from CNN, LSTM, and CNN+LSTM, respectively.

(Continued)

TABLE 3.2 (*Continued*)
Deep Learning Methods for Automated EEG Diagnosis of Alcoholism

Study	Dataset	EEG preprocessing	Classification algorithm	Performance
Li et al., 2022 (63)	UCI KDD, 122 subjects (77 alcoholic and 45 normal)	standard deviation standardization, DWT	DWT-CNN-B-LSTM architecture	accuracy: 99.32%, precision: 99.01%, recall: 98.87%, F1-score: 98.93%

UCI KDD: University of California, Irvine Knowledge Discovery in Database, SVM: support vector machine, KNN: K-nearest neighbor, LVQ: learning vector quantization, RNN: recurrent neural network, LSTM: long short-term memory, B-LSTM: bidirectional long short-term memory, CNN: convolutional neural network, DWT: discrete wavelet transformation, PCA: principal component analysis, ICA: independent component analysis, ATTN: attention mechanism, FFT: fast Fourier transform

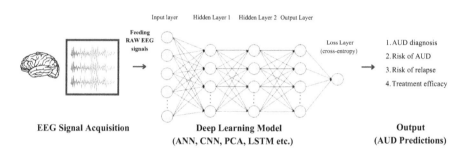

FIGURE 3.3 Workflow of deep learning methods for EEG-based automatic detection of alcoholism.

Similar to machine learning, researchers have also applied time–frequency analysis and relevant statistical analysis, e.g., FFT, discrete/continuous wavelet transformation, principal component analysis, and ICA, in the preprocessing of EEG data with deep learning (25, 59, 61–63). Compared with machine learning, the preprocessing of EEG data in deep learning can be simplified because the features are automatically extracted (Figure 3.3). Mukhtar et al. adopted no preprocessing techniques except for a standard normalization process to avoid wide dynamic ranges in EEG signals (37). Kumari et al. directly used raw EEG data as the model input (60). New methods for EEG preprocessing have also been proposed. Fayyaz et al. integrated machine learning and deep learning, using peak visualization in preprocessing to select the peaks for feature extraction (57).

3.5.2 Classification: Algorithms and Performance

Compared with conventional machine learning, deep learning can detect more underlying features automatically. The feature extraction, selection, and classification are often integrated and adaptive to different EEG signals. Despite the demand for high computational power, deep learning models provide the potential for precise medicine based on large datasets of numerous subjects/cohorts.

The three most commonly utilized deep learning techniques are artificial neural networks (ANNs), CNNs, and recurrent neural networks (RNNs). ANNs are the simplest deep learning algorithm, comprising a group of neurons in each layer that only process the input in the forward direction. CNNs, meanwhile, have garnered attention for their ability to identify images accurately by using convolutions in their architecture, and RNNs are time-series-based algorithms that receive information from inputs continually; they have inbuilt memory cells to retain information from previous outputs. CNNs are effective for analyzing spatial data, while RNN are suitable for temporal data. Mukhtar et al. achieved 98% accuracy in detecting alcoholism using CNN (37). Shen et al. 2022 (25) compared 2D and 3D CNNs in detecting alcoholism and achieved accuracy of 96.25 ± 3.11% using 3D CNN.

Long short-term memory (LSTM) networks are a type of RNN capable of learning order dependence in sequence prediction that appeared in recent works on EEG-based alcoholism detection. Farsi et al. compared the performance of two approaches, i.e., ANN using PCA-derived EEG features and LSTM using raw EEG data. The average accuracy of 10-fold cross-validation was 86% for PCA-ANN and 90% for LSTM (62). Fayyaz et al. used LSTM for feature extraction and SVM to classify normal and alcoholic subjects (57). The developed model achieved an average classification accuracy of 90%, outperforming a control CNN trained on the same data that achieved an accuracy of 88% (57).

Rahman et al. developed a bidirectional LSTM (B-LSTM) and found that it outperformed some other AI algorithms, i.e., learning vector quantization (LVQ), KNN, RNN, and CNN, in detecting alcoholism, with accuracy of 95% (61). However, LSTM does not always outperform CNN. Kumari et al. compared the performance of CNN, LSTM, and a combined framework of both CNN and LSTM in detecting alcoholism. The average classification accuracies were 92.77%, 89%, and 91% along with error rates of 7.5%, 11.90%, and 8.7% from CNN, LSTM, and CNN+LSTM, respectively (60).

Researchers have also developed integrated frameworks that combine different deep learning models. For instance, Li et al. developed a deep learning architecture based on a CNN and a bidirectional LSTM and achieved accuracy of 99.32%, outperforming most previously developed models (63). Neeraj et al. proposed a deep learning architecture that used a combination of fast FFT, CNN, LSTM, and attention mechanism for extracting spatiotemporal features from multichannel EEG signals, with an outstanding accuracy of 98.83% (59).

3.6 DISCUSSION: STRENGTHS AND LIMITATIONS OF CURRENT APPROACHES

There has been recent, rapid development of AI-assisted automatic detection of alcoholism. Overall, current machine learning and deep learning models showed high accuracy in classifying alcoholic and non-alcoholic EEG signals. AI in this field promises to drastically revolutionize the management of AUD by deepening the understanding of the neuropsychiatric effects of alcohol on the human brain, enabling anonymous and highly accurate diagnostic of AUD, assessing treatment response, and predicting the risk of relapse. New EEG features relevant to alcoholism were discovered using advanced signal processing. The integration of preprocessing, feature extraction, feature selection, and classification laid the groundwork for the automatic detection of alcoholism/AUD (12).

Numerous obstacles and restrictions must be overcome before these algorithms can be completely relied upon in clinical settings. With dependence on the UCI KDD data, small sample size is a major limitation of existing AI-assisted alcoholism detection methods and significantly affects the reliability of the results (64). Considering the low sample size for validation, the high performance of some AI models needs to be considered carefully. The lack of proper EEG datasets also limited the exploration of alcoholism-induced changes in EEG measured at different physiological conditions. In addition, current AI methods focus on the detection of alcoholism without considering other relevant clinical needs. Other challenges in clinical practice include the lack of a gold standard, reliance on self-reported data, potential biases and incompleteness in data sources, and the interpretability of "black box" AI models.

3.7 FUTURE DIRECTIONS

In future research, new EEG datasets will play a key role. Multicenter large datasets can be collected to cover different cohorts and the subjects in different treatment

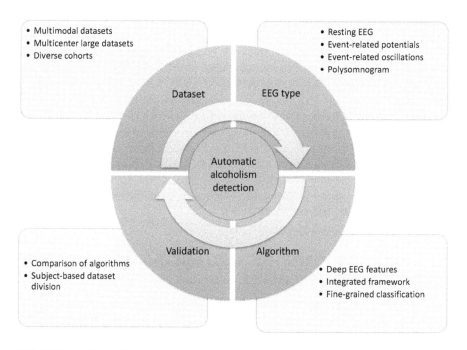

FIGURE 3.4 Future directions of AI-enhanced automatic detection of alcoholism.

strategies. The integration with other types of data (e.g., fMRI, fNIRS, medical history) will generate multimodal datasets to improve the reliability of classification. Different types of EEG signals such as resting EEG, event-related potentials, event-related oscillations, and polysomnography, have been reported as potential tools for alcoholism detection (12, 64, 65). The variations in experimental paradigms and more advanced EEG data analysis methods have been considered to explore underlying neurocognitive and psychological mechanisms of alcoholism (64).

Regarding the algorithms, using advanced signal processing methods, more alcoholism-relevant EEG features can be discovered. The comprehensive comparison of different AI algorithms and the integrated frameworks that combine different algorithms will improve the performance of classification and make it adaptive to different cohorts/EEG datasets. Based on more new datasets, existing AI algorithms can be further validated where subject-based division of training, validation, and testing sets (i.e., EEG data of different subjects will be randomly distributed in different sets) will improve the reliability of the algorithms. The fine-grained classification methods can be developed to achieve the stratification of alcoholism (e.g., abstainers, alcoholics, and normal subjects) and the evaluation of new medications or behavioral interventions (12).

REFERENCES

1. Organization WH. *Global status report on alcohol and health 2018*. Geneva, Switzerland: World Health Organization; 2019.
2. Organization WH. *Global health risks: Mortality and burden of disease attributable to selected major risks*. Geneva, Switzerland: World Health Organization; 2009.
3. Castillo-Carniglia A, Keyes KM, Hasin DS, Cerdá M. Psychiatric comorbidities in alcohol use disorder. The Lancet Psychiatry. 2019;6(12):1068–1080.
4. Grant BF, Goldstein RB, Saha TD, Chou SP, Jung J, Zhang H, et al. Epidemiology of DSM-5 alcohol use disorder: Results from the national epidemiologic survey on alcohol and related conditions III. JAMA Psychiatry. 2015;72(8):757–766.
5. Hendriks HFJ. Alcohol and human health: What is the evidence? Annual Review of Food Science and Technology. 2020;11(1):1–21.
6. Day E, Rudd JHF. Alcohol use disorders and the heart. Addiction. 2019;114(9):1670–1678.
7. Levit JD, Meyers JL, Georgakopoulos P, Pato MT. Risk for alcohol use problems in severe mental illness: Interactions with sex and racial/ethnic minority status. Journal of Affective Disorders. 2023;325:329–336.
8. Ziring DJ, Adler AG. Alcoholism: Are you missing the diagnosis? Postgraduate Medicine. 1991;89(5):139–145.
9. Beniczky S, Schomer DL. Electroencephalography: Basic biophysical and technological aspects important for clinical applications. Epileptic Disorders. 2020;22(6):697–715.
10. Coutin-Churchman P, Moreno R, Añez Y, Vergara F. Clinical correlates of quantitative EEG alterations in alcoholic patients. Clinical Neurophysiology. 2006;117(4):740–751.
11. Kamarajan C. Chapter 13—Brain electrophysiological signatures in human alcoholism and risk. In: Preedy VR, editor. *Neuroscience of alcohol*. Amsterdam, Netherlands: Academic Press; 2019. pp. 119–130. https://doi.org/10.1016/B978-0-12-813125-1.00013-1
12. Mumtaz W, Vuong PL, Malik AS, Rashid RBA. A review on EEG-based methods for screening and diagnosing alcohol use disorder. Cognitive Neurodynamics. 2018;12(2):141–156.
13. Acharya UR, S V, Bhat S, Adeli H, Adeli A. Computer-aided diagnosis of alcoholism-related EEG signals. Epilepsy & Behavior. 2014;41:257–263.

14. Newson JJ, Thiagarajan TC. EEG frequency bands in psychiatric disorders: A review of resting state studies. Frontiers in Human Neuroscience. 2019;12.
15. Rangaswamy M, Porjesz B, Chorlian DB, Choi K, Jones KA, Wang K, et al. Theta power in the EEG of alcoholics. Alcohol: Clinical and Experimental Research. 2003;27(4):607–615.
16. Saletu-Zyhlarz GM, Arnold O, Anderer P, Oberndorfer S, Walter H, Lesch OM, et al. Differences in brain function between relapsing and abstaining alcohol-dependent patients, evaluated by EEG mapping. Alcohol and Alcoholism. 2004;39(3):233–240.
17. Campanella S, Petit G, Maurage P, Kornreich C, Verbanck P, Noël X. Chronic alcoholism: Insights from neurophysiology. Neurophysiologie Clinique/Clinical Neurophysiology. 2009;39(4):191–207.
18. Rangaswamy M, Porjesz B. Chapter 23—Understanding alcohol use disorders with neuroelectrophysiology. In: Sullivan EV, Pfefferbaum A, editors. *Handbook of clinical neurology*. 125: Amsterdam, Netherlands: Elsevier; 2014. pp. 383–414.
19. Hodgkinson CA, Enoch M-A, Srivastava V, Cummins-Oman JS, Ferrier C, Iarikova P, et al. Genome-wide association identifies candidate genes that influence the human electroencephalogram. Proceedings of the National Academy of Sciences. 2010;107(19):8695–8700.
20. Mumtaz W, Vuong PL, Xia L, Malik AS, Rashid RBA. Automatic diagnosis of alcohol use disorder using EEG features. Knowledge-Based Systems. 2016;105:48–59.
21. Rangaswamy M, Porjesz B, Chorlian DB, Wang K, Jones KA, Bauer LO, et al. Beta power in the EEG of alcoholics. Biological Psychiatry. 2002;52(8):831–842.
22. de Bruin EA, Stam CJ, Bijl S, Verbaten MN, Kenemans JL. Moderate-to-heavy alcohol intake is associated with differences in synchronization of brain activity during rest and mental rehearsal. International Journal of Psychophysiology. 2006;60(3):304–314.
23. Bavkar S, Iyer B, Deosarkar S. Rapid screening of alcoholism: An EEG based optimal channel selection approach. IEEE Access. 2019;7:99670–99682.
24. Kinreich S, Meyers JL, Maron-Katz A, Kamarajan C, Pandey AK, Chorlian DB, et al. Predicting risk for alcohol use disorder using longitudinal data with multimodal biomarkers and family history: A machine learning study. Molecular Psychiatry. 2021;26(4):1133–1141.
25. Shen M, Wen P, Song B, Li Y. Detection of alcoholic EEG signals based on whole brain connectivity and convolution neural networks. Biomedical Signal Processing and Control. 2023;79:104242.
26. Winterer G, Klöppel B, Heinz A, Ziller M, Dufeu P, Schmidt LG, et al. Quantitative EEG (QEEG) predicts relapse in patients with chronic alcoholism and points to a frontally pronounced cerebral disturbance. Psychiatry Research. 1998;78(1):101–113.
27. Bauer LO. Predicting relapse to alcohol and drug abuse via quantitative electroencephalography. Neuropsychopharmacology. 2001;25(3):332–340.
28. Lourenço C, Tjepkema-Cloostermans MC, Teixeira LF, van Putten MJAM, editors. Deep learning for interictal epileptiform discharge detection from scalp EEG recordings. XV Mediterranean Conference on Medical and Biological Engineering and Computing—MEDICON 2019: Cham: Springer International Publishing; 2020.
29. Flanary J, Daly S, Bakker C, Herman AB, Park MC, McGovern R, et al. Reliability of visual review of intracranial electroencephalogram in identifying the seizure onset zone: A systematic review and implications for the accuracy of automated methods. Epilepsia. 2023;64(1):6–16.
30. Suzuki Y, Suzuki M, Shigenobu K, Shinosaki K, Aoki Y, Kikuchi H, et al. A prospective multicenter validation study of a machine learning algorithm classifier on quantitative electroencephalogram for differentiating between dementia with Lewy bodies and Alzheimer's dementia. PLOS ONE. 2022;17(3):e0265484.

31. Hosseini M-P, Hosseini A, Ahi K. A review on machine learning for EEG signal processing in bioengineering. IEEE Reviews in Biomedical Engineering. 2020;14:204–218.
32. Khademi Z, Ebrahimi F, Kordy HM. A review of critical challenges in MI-BCI: From conventional to deep learning methods. Journal of Neuroscience Methods. 2023;383:109736.
33. Zhang H, Silva FH, Ohata EF, Medeiros AG, Rebouças Filho PP. Bi-dimensional approach based on transfer learning for alcoholism pre-disposition classification via EEG signals. Frontiers in Human Neuroscience. 2020;14:365.
34. Bae Y, Yoo BW, Lee JC, Kim HC. Automated network analysis to measure brain effective connectivity estimated from EEG data of patients with alcoholism. Physiological Measurement. 2017;38(5):759.
35. Anuragi A, Sisodia DS. Empirical wavelet transform based automated alcoholism detecting using EEG signal features. Biomedical Signal Processing and Control. 2020;57:101777.
36. Buriro AB, Ahmed B, Baloch G, Ahmed J, Shoorangiz R, Weddell SJ, et al. Classification of alcoholic EEG signals using wavelet scattering transform-based features. Computers in Biology and Medicine. 2021;139:104969.
37. Mukhtar H, Qaisar SM, Zaguia A. Deep convolutional neural network regularization for alcoholism detection using EEG signals. Sensors. 2021;21(16):5456.
38. Bavkar S, Iyer B, Deosarkar S, editors. Detection of alcoholism: An EEG hybrid features and ensemble subspace K-NN based approach. In *Distributed computing and internet technology*: Cham: Springer International Publishing; 2019.
39. Sadiq MT, Siuly S, Ur Rehman A, Wang H, editors. *Auto-correlation based feature extraction approach for EEG alcoholism identification. Health information science*: Cham: Springer International Publishing; 2021.
40. Rodrigues JDC, Filho PPR, Peixoto E, Arun Kumar N, de Albuquerque VHC. Classification of EEG signals to detect alcoholism using machine learning techniques. Pattern Recognition Letters. 2019;125:140–149.
41. Anuragi A, Sisodia DS, Pachori RB. Automated alcoholism detection using fourier-bessel series expansion based empirical wavelet transform. IEEE Sensors Journal. 2020;20(9):4914–4924.
42. Sadiq MT, Akbari H, Siuly S, Li Y, Wen P, editors. *Fractional Fourier transform aided computerized framework for alcoholism identification in EEG. Health information science*: Cham: Springer Nature Switzerland; 2022.
43. Bavkar S, Iyer B, Deosarkar S. Optimal EEG channels selection for alcoholism screening using EMD domain statistical features and harmony search algorithm. Biocybernetics and Biomedical Engineering. 2021;41(1):83–96.
44. Agarwal S, Zubair M. Classification of alcoholic and non-alcoholic EEG signals based on sliding-SSA and independent component analysis. IEEE Sensors Journal. 2021;21(23):26198–26206.
45. Salankar N, Qaisar SM, Pławiak P, Tadeusiewicz R, Hammad M. EEG based alcoholism detection by oscillatory modes decomposition second order difference plots and machine learning. Biocybernetics and Biomedical Engineering. 2022;42(1):173–186.
46. Acharya UR, Sree SV, Chattopadhyay S, Suri JS. Automated diagnosis of normal and alcoholic EEG signals. International Journal of Neural Systems. 2012;22(3):1250011.
47. Zhu G, Li Y, Wen P, Wang S. Analysis of alcoholic EEG signals based on horizontal visibility graph entropy. Brain Informatics. 2014;1(1):19–25.
48. Michael A, Mirza KAH, Mukundan CR, Channabasavanna SM. Interhemispheric electroencephalographic coherence as a biological marker in alcoholism. Acta Psychiatrica Scandinavica. 1993;87(3):213–217.
49. Tcheslavski GV, Gonen FF. Alcoholism-related alterations in spectrum, coherence, and phase synchrony of topical electroencephalogram. Computers in Biology and Medicine. 2012;42(4):394–401.

50. Prabhakar SK, Rajaguru H. Alcoholic EEG signal classification with correlation dimension based distance metrics approach and modified Adaboost classification. Heliyon. 2020;6(12):e05689.

51. Wang Q, Loh JM, He X, Wang Y. A latent state space model for estimating brain dynamics from electroencephalogram (EEG) data. Biometrics. 2022;n/a(n/a).

52. Sadiq MT, Akbari H, Siuly S, Li Y, Wen P. Alcoholic EEG signals recognition based on phase space dynamic and geometrical features. Chaos, Solitons & Fractals. 2022;158:112036.

53. Pain S, Roy S, Sarma M, Samanta D. Detection of alcoholism by combining EEG local activations with brain connectivity features and Graph Neural Network. Biomedical Signal Processing and Control. 2023;85:104851.

54. Siuly S, Bajaj V, Sengur A, Zhang Y. An advanced analysis system for identifying alcoholic brain state through EEG signals. International Journal of Automation and Computing. 2019;16(6):737–747.

55. Rajaguru H, Vigneshkumar A, Gowri Shankar M. Alcoholic EEG signal classification using multi-heuristic classifiers with stochastic gradient descent technique for tuning the hyperparameters. IETE Journal of Research. 2023:1–16.

56. Kumari N, Anwar S, Bhattacharjee V. Correlation and relief attribute rank-based feature selection methods for detection of alcoholic disorder using electroencephalogram signals. IETE Journal of Research. 2022;68(5):3816–3828.

57. Fayyaz A, Maqbool M, Saeed M. Classifying alcoholics and control patients using deep learning and peak visualization method. Proceedings of the 3rd International Conference on Vision, Image and Signal Processing: Vancouver, BC, Canada: Association for Computing Machinery; 2020. p. Article 60.

58. Dong F, Wang X. A classifier for multi-dimensional datasets based on Bayesian multiple kernel grouping learning. Journal of Statistical Computation and Simulation. 2019;89(11):2151–2174.

59. Neeraj, Singhal V, Mathew J, Behera RK. Detection of alcoholism using EEG signals and a CNN-LSTM-ATTN network. Computers in Biology and Medicine. 2021;138:104940.

60. Kumari N, Anwar S, Bhattacharjee V. A deep learning-based approach for accurate diagnosis of alcohol usage severity using EEG signals. IETE Journal of Research. 2022:1–15.

61. Rahman S, Sharma T, Mahmud M, editors. *Improving alcoholism diagnosis: Comparing instance-based classifiers against neural networks for classifying EEG signal. Brain informatics*: Cham: Springer International Publishing; 2020.

62. Farsi L, Siuly S, Kabir E, Wang H. Classification of alcoholic EEG signals using a deep learning method. IEEE Sensors Journal. 2020;21(3):3552–3560.

63. Li H, Wu L. EEG classification of normal and alcoholic by deep learning. Brain Sciences. 2022;12(6):778.

64. Zhang H, Yao J, Xu C, Wang C. Targeting electroencephalography for alcohol dependence: A narrative review. CNS Neuroscience & Therapeutics. 2023;29(5):1205–1212.

65. Jurado-Barba R, Sion A, Martínez-Maldonado A, Domínguez-Centeno I, Prieto-Montalvo J, Navarrete F, et al. Neuropsychophysiological measures of alcohol dependence: Can we use EEG in the clinical assessment? Frontiers in Psychiatry. 2020;11:676.

4 Empirical Wavelet Transform and Gradient Boosted Learners for Automated Classification of Epileptic Seizures from EEG Signals

Mohd Faizan Bari, Dilip Singh Sisodia, and Arti Anuragi

4.1 INTRODUCTION

After Alzheimer's, epilepsy is the second most prevalent neurological disorder. With around 50 million individuals affected by epilepsy worldwide [1], it can impact individuals of all ages, emphasizing the significance of diagnosing this condition. Patients with epilepsy experience seizures that result in uncontrolled movements or loss of consciousness, often leading to severe injury or even death. The electrical activity of the brain is represented by the electroencephalogram (EEG) signal, which displays the voltage amplitude over a period of time. This recorded data is considered time-series data and can be utilized for an epilepsy diagnosis. The EEG signal is collected by placing electrodes on a person's scalp. However, manually inspecting these signals is a complicated task that requires experts and can be time-consuming. Moreover, human error can cause inaccuracies in the analysis.

Several automated methods have emerged in recent years to differentiate between epileptic and nonepileptic patients. Initially, during the early stages of seizure detection, Fourier transform techniques like discrete Fourier transform and fast-Fourier transform (FFT) was employed, assuming EEG signals to be stationary. However, it was later discovered that EEG signals are nonstationary [2]. Thus, a time–frequency representation of the EEG signal is required, which can be obtained using wavelet transform. A wavelet is a wave-like oscillation that has a limited duration and an average value of zero. Wavelet transform is a mapping of signals in the time-frequency domain, using functions that are localized in both real and Fourier space [3]. The wavelet transform is expressed by the following equation Eq. (1):

DOI: 10.1201/9781003479970-4

$$F(a,b) = \int_{-\infty}^{\infty} f(x) \Psi^*_{(a,b)}(x)\, dt \qquad\qquad (1)$$

where * is a complex conjugate symbol and function Ψ is some function that can be chosen arbitrarily provided that it obeys certain rules.

Various wavelet transform methods have been employed to decompose EEG signals, including discrete wavelet transform (DWT), tunable Q wavelet transform (TQWT) [4], flexible analytical wavelet transform [5], and empirical wavelet transform (EWT) [6]. Additionally, numerous techniques that employed wavelet transform and machine learning classifiers have been developed to distinguish between seizure and non-seizure EEG signals [5–14]. In their study, Gupta et al. [7] introduced a new filtering method that combines TQWT and empirical mode decomposition (EMD) to decompose EEG signals; they extracted information potential features from each resulting sub-band signal and utilized these features for training a least square support vector machine (LS-SVM) classifier. This approach achieved an impressive classification accuracy of 97%. In [8], Saxena et al. applied EMD and EWT methods to decompose EEG signals and computed bandwidth using amplitude modulation and frequency modulation. They utilized the resulting feature vector to classify the signals using LS-SVM with a polynomial kernel function. They achieved an accuracy of 98.4% in their classification task. In another study, Bhattacharyya et al. [9] developed a new approach to compute multiscale k-nearest neighbors (KNN) entropy from sub-band signals generated by TQWT and then use the extracted features in a support vector machine (SVM) classifier that utilized wrapper-based feature selection. This method demonstrated perfect accuracy of 100% in differentiating normal (eyes-open and eyes-closed) and seizure EEG signals.

In another study, Sharaf et al. [10] used TQWT to classify EEG signal-based epilepsy. After that, features such as chaotic, statistical, and power spectrum were extracted from each sub-band signal. Then firefly optimization was employed to determine the most significant features and passed to a random forest classifier for classifying seizures and seizure-free signals and achieved accuracy of 99%. In their study, Orhan et al. [11] developed a framework that involved obtaining sub-band signals using DWT, clustering the wavelet coefficient using the k-means algorithm, and computing probability distributions based on the distribution of wavelet coefficients to the clusters. These probability distributions were then used as inputs to a multilayer perceptron neural network model, which achieved a classification accuracy of 96.67% for distinguishing between healthy subjects, epilepsy subjects with seizure-free intervals, and epilepsy subjects with seizures.

In another study, Hadj-Youcef et al. [12] utilized DWT to decompose the EEG signal, after which they extracted six statistical features that produced high-dimensional feature vectors. To reduce this dimensionality, they employed principal component analysis, which resulted in a reduced feature vector that was then passed to an SVM classifier for classification. Their approach achieved accuracy of 98% in the classification task. Authors of another study [13] utilized DWT to decompose EEG signals and then extracted mean, average power, standard deviation, and the ratio of a mean of adjacent sub-band signals. These features were subsequently passed through an artificial neural network for classification, achieving classification accuracy of

93.2%. In their work, [14], Nicoletta and Georgiou presented a tool for detecting epilepsy by using permutation energy as a feature. The obtained features were then passed to an SVM classifier to classify epileptic-seizure and healthy EEG signals, achieving accuracy of 86%. Akbari et al. [15] presented a methodology for detecting epileptic seizures that involved using EWT to obtain EEG rhythms. They then computed reconstructed phase spaces from the obtained rhythms, from which they calculated 95% of the elliptical pattern for features that they fed to a KNN classifier to classify seizure or non-seizure EEG signals. Their approach achieved accuracy of 98%. Jindal et al. [16] utilized flexible analytical wavelet transform to decompose EEG signals into sub-band signals; extracted mean, kurtosis, and skewness from wavelet coefficients; and used LS-SVM, KNN, and random forest classifiers for classification.

The literature reviewed suggests that while numerous studies have aimed to achieve accurate classification of epileptic-seizure EEG signals, there is still room for improvement. Hence, we propose with this paper using EWT to decompose EEG signals into five sub-band signals. The primary advantage of EWT over other wavelet transform methods is that it offers improved resolution, and the filter band formed is very tight, resulting in no loss of necessary or addition of unwanted information. Additionally, EWT has been utilized in previous studies such as [17] for denoising ECG signals and has demonstrated its effectiveness.

The proposed method involves three steps. First, the EEG signal is decomposed into five sub-bands with four boundary values using EWT from a single channel signal of each set of given datasets [18]. After the sub-bands are obtained from the EEG signals, mean, maximum, minimum, standard deviation, log entropy, Shannon entropy, skewness, and kurtosis are extracted from each sub-band. These features are then input into a machine learning classifier using gradient boosting to distinguish between seizure and non-seizure patients.

4.2   MATERIAL AND METHODS

The proposed framework employs the EWT method for analyzing EEG signals by decomposing them into sub-band signals. After decomposing each signal, statistical and entropy features are extracted and used as input for the machine learning classifier to classify seizure or seizure-free EEG signals. In this study, we utilized three different classifiers from the Ensemble learning model, which are known to produce better results than commonly used classifiers. Specifically, we used XG-BOOST, LGBOOST, and CAT-BOOST. The diagram of the proposed framework is depicted in Figure 4.1.

4.2.1   DATASET DESCRIPTION

For this paper, we utilized a dataset obtained from the University of Bonn in Germany that includes five different sets (A, B, C, D, and E) of 100 single-channel EEG signals [19]. Each signal has a duration of 23.6 seconds and a sampling rate of 173.61Hz, with 4097 amplitude samples per channel. Set A comprises EEG recordings of five healthy patients with open eyes, while Set B includes EEG recordings of healthy

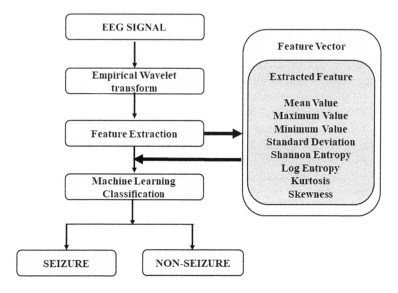

FIGURE 4.1   Workflow diagram of the proposed framework for seizure and non-seizure EEG signal classification.

TABLE 4.1
Description of the Considered Dataset in This Study

Sets	Number of signals	Class
Set A	100	Normal
Set B	100	Normal
Set C	100	Seizure-free
Set D	100	Seizure-free
Set E	100	Seizure

patients with closed eyes. Set C contains EEG recordings from the opposite side of the brain during seizure-free intervals, while Set D comprises EEG recordings from the epileptogenic zone during seizure-free periods. Finally, Set E contains EEG recordings during a seizure. The paper includes Table 4.1 with a summary of the dataset and Figure 4.2 with a plotted sample from each set showing the amplitude versus the sample number.

4.2.2   EMPIRICAL WAVELET TRANSFORM

EWT involves decomposing signals by generating an adaptive filter bank that matches the spectrum of the input signal [20]. To establish the appropriate bandpass for the adaptive filter bank, the spectrum is segmented in a suitable manner. The

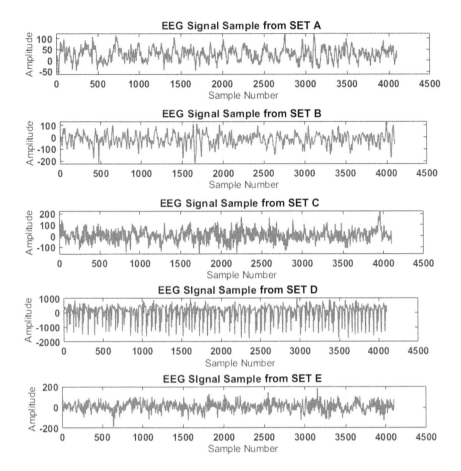

FIGURE 4.2   EEG signal sample from each studied dataset.

EWT toolbox provides several methods for accurately segmenting the spectrum. In the Fourier domain, filters for the scaling function $\Phi(w_f)$ and wavelet functions $\Psi(w_f)$ are produced using Little Wood-Paley and Meyer wavelets.

Let's assume that the Fourier support interval $[0, \pi]$ has been segmented into N-contiguous segments. Let's denote w_n to be limited between each segment (where $w_0 = 0$ and $w_n = \pi$). Each segment is denoted by $A_n = [w_{n-1}, w_n]$. Then, it is easy to see that each segment is centered around each w_n and the transition phase is of width $2\tau_n$. In EWT, the empirical scaling $\Phi(n)$ and wavelet function $\Psi(n)$ are defined as shown in Eq. (2) and Eq. (3), respectively.

$$\Phi(n) = \begin{cases} 1 & \text{if } |w| \le w_n - \tau_n \\ \cos\left[\dfrac{\pi}{2}\beta\left(\dfrac{1}{2\tau_n}\left(|w| - w_n + \tau_n\right)\right)\right] & \text{if } w_n - \tau_n \le |w| \le w_n + \tau_n \\ 0 & \text{otherwise} \end{cases} \quad (2)$$

$$\Psi(n) = \begin{cases} 1 & \text{if } w_n + \tau_n \le |w| \le w_{n+1} - \tau_{n+1} \\ \cos\left[\dfrac{\pi}{2}\beta\left(\dfrac{1}{2\tau_{n+1}}\left(|w| - w_{n+1} + \tau_{n+1}\right)\right)\right] & \text{if } w_{n+1} - \tau_{n+1} \le |w| \le w_{n+1} + \tau_{n+1} \\ \sin\left[\dfrac{\pi}{2}\beta\left(\dfrac{1}{2\tau_n}\left(|w| - w_n + \tau_n\right)\right)\right] & \text{if } w_n - \tau_n \le |w| \le w_n + \tau_n \end{cases} \tag{3}$$

where $\beta(w, x)$ is an arbitrary function and is mathematically expressed as shown in Eq. (4):

$$\beta(x) = \begin{cases} 0 & \text{if } x \le 0 \text{ and } \beta(x) + \beta(1-x) = 1 \quad \forall x \in [0,1] \\ 1 & \text{if } x \ge 1 \end{cases} \tag{4}$$

Concerning the choice of τ_n, several options are possible. The simplest is to choose τ_n proportional to $w_n : \tau_n = \gamma w_n$ where $0 < \gamma < 1$.

The EWT can be defined in a manner similar to the classical wavelet transform, where detail coefficients are calculated through inner products with empirical wavelets. The expression for the detail coefficient is given in Eq. (5):

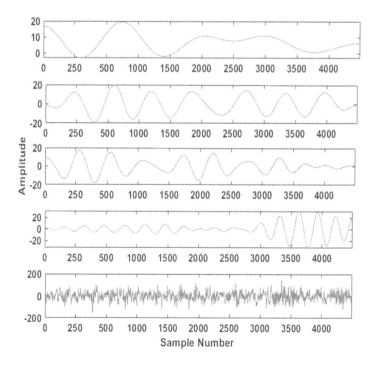

FIGURE 4.3 Decomposed EEG signal of a normal subject from SET A.

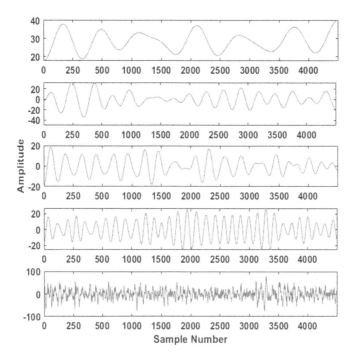

FIGURE 4.4   Decomposed EEG signal of an epileptic subject from SET E.

$$W_f^e (n,t) = \int f(\tau) \overline{\Psi_n(\tau - t)}\, d\tau \tag{5}$$

Similarly, the approximation coefficient is produced by the inner product with the scaling function:

$$W_f^e (0,t) = \int f(\tau) \overline{\Phi_n(\tau - t)}\, d\tau \tag{6}$$

The EEG signals from Set A and Set E were decomposed into five sub-band signals using EWT in this study. The resulting decomposed EEG signals from each set can be seen in Figure 4.3 and Figure 4.4.

4.2.3   FEATURE EXTRACTION

The primary objective of feature extraction from raw data is to decrease the size of the input feature vector, as well as to eliminate any redundant information present in EEG signals. In this study, we extracted statistical and entropy features from the sub-band signals of EEG signals, which we obtained after applying the EWT. We computed a total of eight features from each sub-band signal: maximum, minimum, mean value, standard deviation, log entropy, Shannon entropy, skewness, and kurtosis. Table 4.2 offers a concise description of the features that we extracted from decomposed EEG signals, where N represents the total number of sample points from the signal.

TABLE 4.2

A Detailed Description of Extracted Features

S. no	Features	Description	Formula
1	Maximum	The highest amplitude of the sub-band	Max (sub-band)
2	Minimum	The lowest amplitude of the sub-band	Min (sub-band)
3	Mean	The ratio of the sum of observations to the total number of observations	$\mu = \dfrac{1}{N}\sum_{i=1}^{N} D_i \quad i = 1,2,\dots N$
4	Standard deviation	The dispersion of the dataset relative to its mean	$\sigma = \sqrt{\dfrac{1}{N-1}\sum_{i=1}^{N}(D_i - \mu)^2} \quad i = 1,2,\dots.N$
5	Kurtosis	A measure of the peakedness of data distribution	$\text{Kurtosis} = \dfrac{\sum_{i=1}^{N}\dfrac{X_i - \mu}{N}}{\sigma^4} \quad i = 1,2,\dots N$
6	Skewness	A measure of the asymmetry of data distribution	$Skewness = \dfrac{1}{N}\sum_{i=1}^{N}\left(\dfrac{D_i - \mu}{\sigma}\right)^4 - 3 \quad i = 1,2,\dots N$
7	Log entropy	The degree of complexities in an EEG signal	$H_{logEn}(x) = -\sum_{i=0}^{N-1}\left(\log_2\left(p_i(x)\right)\right)^2 \quad i = 1,2,\dots N$
8	Shannon entropy	The uncertain information in an EEG signal	$H_{SE}(x) = -\sum_{i=0}^{N-1}\left(p_i(x)\right)^2\left(\log_2\left(p_i(x)\right)\right)^2 \quad i = 1,2\dots N$

After extracting these features from both seizure and seizure-free EEG signals, which consisted of 100 signals each, we utilized gradient-boosting classifiers to differentiate between the two based on whether a seizure was present or not. The obtained feature vector is in the matrix of 200 (no. of EEG signals) × 40 (total extracted features) dimensions.

4.2.4 GRADIENT-BOOSTED CLASSIFIERS

For the paper, we employed ensemble learning to categorize EEG signals as either seizure or non-seizure. Ensemble learning is a method of solving a computational intelligence problem by generating and combining multiple models, such as classifiers, in a strategic manner; the fundamental concept behind ensemble learning is to enhance the weak learner by combining two or more of them, with the decision being regarded as a weak learner. There are two categories of ensemble learning: boosting and bagging [21]. In the boosting algorithm, models are employed in a series combination, where the output from one model serves as the input for the next model, and so on. Boosting is further divided into two types: (a) adaptive boosting (AdaBoost) and (b) gradient boosting (GB).

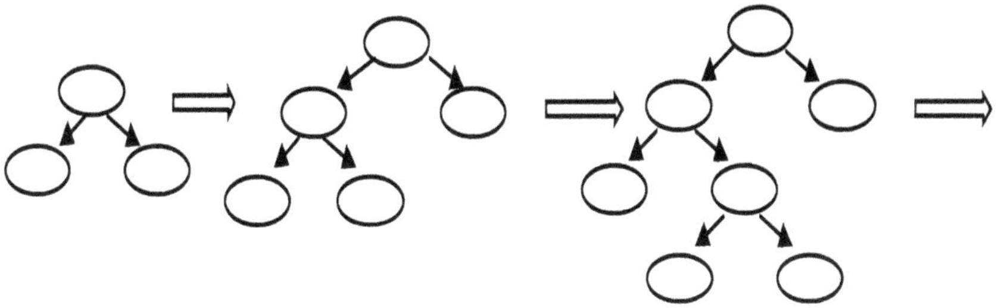

FIGURE 4.5 Leaf-wise split in light GBM.

The AdaBoost algorithm improves classification accuracy by assigning a higher weight to incorrectly predicted samples from the previous model, continuing to do so until the last model in the combination. On the other hand, in GB, instead of changing weights for every incorrectly classified observation, as in AdaBoost, GB attempts to fit new predictors to the residual error produced by the previous model. It utilizes the concept of gradient descent to calculate the residual error. Gradient boosting is divided into three techniques: (a) XGBoost, (b) light boost, and (c) categorical boost. In this study, we utilized all three techniques to categorize EEG signals as either seizure or seizure-free.

XGBoost is a distributed gradient boosting library that is optimized for high efficiency, flexibility, and portability. It implements machine learning algorithms based on the gradient boosting framework. XGBoost [22] is an optimized gradient-boosting algorithm that utilizes parallel processing, tree pruning, handling of missing data, and regularization to prevent overfitting. Compared with standard gradient boosting algorithms, the training time of XGBoost is considerably reduced. To achieve better performance parameters, certain parameters must be tuned.

The light gradient boosting model utilizes a unique method called gradient-based side sampling to filter out data instances for identifying split values. In this algorithm, trees are split leaf-wise with the best fit, unlike other boosting techniques that split the tree by depth or level. This approach provides several benefits, including faster training speed compared with XGBoost, lower memory usage, improved accuracy, and support for GPU learning. It has the following parameters: max_depth, min_data_in_leaf, learning_rate, feature_fraction, early_stopping_round, and num_leaves [23].

Cat boost library [24] was developed at Yandex. It is the latest gradient-boosting library, and it outperforms both XGBoost and light GBM. It is a new open-source gradient-boosting library that successfully handles categorical features. The tuning parameters of this boosting technique are cat_features, learning_rate, n_estimator, max_depth, and one_hot_max_size.

4.3 RESULT AND DISCUSSION

In this section, we present the results achieved through the proposed framework. First, we decomposed EEG signals from two sets, namely SET A (normal) and SET E (seizure), into five levels and obtained the five sub-band signals using EWT, as depicted in Figure 4.3 and Figure 4.4, respectively. After decomposition, we extracted eight features (statistical and entropy) for each level, resulting in a total of forty features for each class. These features were then applied to ensemble machine learning classifiers, namely XGBoost, Light Boost, and CatBoost classifiers. For the classifiers in this study, we evaluated three performance parameters, namely accuracy, sensitivity, and specificity, using standard measures and the following mathematical expressions of each considered performance parameter:

$$Accuracy = \frac{TP + TN}{TP + TN + FP + FN} \tag{7}$$

$$Sensitivity = \frac{TP}{TP + FN} \tag{8}$$

$$Specificity = \frac{TN}{TN + FP} \tag{9}$$

TABLE 4.3

Performance Parameters of Machine Learning Classifiers

Author	Decomposition technique	Features	Classifier	Accuracy (%)
Acharya et al. [25]	EMD	Spectral peaks, energy, entropy	CART Decision Tree, C4.5 Decision Tree	CART-94.33 C4.5–95.33
Gupta et al. [7]	TQWT & EMD	Information Potential	LS-SVM	97
Patidar et al. [26]	TQWT	Kraskov entropy-based feature extraction	LS-SVM	97
Thilagaraj et al. [27]	EWT	Maximum, minimum, mean, standard deviation	k-NN and J48 Decision Tree	k-NN-94.2 J48–96.2
Yinxia Liu et al. [28]	DWT	Relative amplitude, standard deviation, fluctuation index, relative energy	SVM	95.33
Saxena et al. [8]	EMD & EWT	Amplitude modulation, frequency modulation	LS-SVM	RBF-95.52 Polynomial-98.40 Linear-95.45
Proposed method	EWT	Statistical and entropy-based features	Cat Boost Light Boost XG-Boost	98.33 96.67 95.45

TABLE 4.4
Comparison of the Proposed Method with Existing Methods

Classifier	Accuracy (%)	Sensitivity (%)	Specificity (%)
XG-BOOST	95.45	100	92.11
Light-GBM	96.67	96.43	96.88
Cat Boost	98.33	96.88	100

where TP in Eq. (7–9) denotes true positive, TN signifies true negative, FP denotes false positive, and FN signifies false negative.

Two sets, namely SET A and SET E, were selected from the provided dataset. SET A includes normal EEG signals, while SET E contained seizure EEG signals in this study for forming the feature vector. To avoid overfitting, we used hold-out cross-validation with a 70–30% split ratio for training and testing, respectively. We split the feature vector into 70–30% for training and testing the models, and the classifiers used were XGBOOST, light GBM, and CatBoost. Table 4.3 shows that the categorical boosting classifier had the highest classification accuracy of 98.33%, sensitivity of 96.88%, and specificity of 100%.

Light gradient boosting was the second best-performing classifier, with accuracy of 96.67%, sensitivity of 96.43%, and specificity of 96.88%. XG-Boost had the lowest performance in the proposed framework, with accuracy of 95.45%, sensitivity of 100%, and specificity of 92.11%. Additionally, Table 4.4 compares the proposed framework's performance with other state-of-the-art methods on the same dataset. Table 4.4 indicates that the proposed framework surpasses other methods and achieves exceptional results, with the CatBoost classifier achieving the highest classification accuracy of 98.33%. These noteworthy results suggest that the proposed framework could aid clinicians in detecting epileptic-seizure EEG signals. However, it should be noted that the proposed framework has limitations, as it was only tested on a small number of samples. To develop a more reliable and generalizable model, we recommend testing the framework on a larger and more diverse dataset.

4.4 CONCLUSIONS

The manual detection of epileptic seizures is a challenging and time-consuming task due to the nonstationary nature of EEG signals. To improve the detection of seizures in EEG signals, this paper proposes a method that decomposes each EEG signal into five sub-bands using the EWT and extracts statistical and entropy features for each sub-band. These features are then inputted into a gradient-boosting classifier model, which uses XG-BOOST, Light Boost, and Cat Boost algorithms to classify seizure and non-seizure EEG signals. The achieved accuracy demonstrates the potential of this model for diagnosing epileptic seizures from EEG signals. To the best of our knowledge, this is the first time that EWT has been used in combination with a gradient-boosting classifier for detecting epileptic seizures. Future work on this topic could involve extracting more features or decomposing the EEG into more or fewer sub-bands using EWT.

4.5 AUTHOR BIOGRAPHIES

Mohd Faizan Bari studied M. Tech in the department of Information Technology of the National Institute of Technology, Raipur, India (NIT), during 2018–19 and completed his B. Tech. from Dhhattisgarh Swami Vivekanand Technical University, Bhillai. His research interests include the signal processing, and machine learning.

Dilip Singh Sisodia is an associate professor in the Department of Computer Science Engineering of NIT, Raipur. Dr. Sisodia has contributed over 125 high-impact articles in reputed journals, conference proceedings, and edited volumes. He also edited four Scopus-indexed research book volumes published by Springer Nature and IGI Global. Dr. Sisodia was included in the World Ranking of Top 2% Scientists/researchers in 2022 & 2023 by a study of scientists from Stanford University (USA) and has been published by Elsevier B.V. He has supervised six Ph.D. theses and nine M. Tech. dissertations. He is a senior member of IEEE and ACM. He received a Ph.D. degree in computer science and engineering from NIT. He earned his M. Tech. and B.E. degrees, respectively, in information technology (with specialization in artificial intelligence) and computer science and engineering from the Rajiv Gandhi Technological University, Bhopal. His research interests include applications of machine learning/soft computing, artificial intelligence, biomedical signal, and image processing.

Arti Anuragi received the B. Tech. degree in electronics and communication from Parul Institute of Technology, Vadodara, Gujarat, India, in 2015, and the M. Tech. degree in information technology from NIT in 2018. Recently, she submitted her Ph.D. thesis in computer science and engineering at NIT. She has published more than 13 papers in international journals and conferences. Her research interests include biomedical signal processing, nonstationary signal processing, and machine learning.

REFERENCES

1. I. Ullah, M. Hussain, E. ul H. Qazi, and H. Aboalsamh, "An automated system for epilepsy detection using EEG brain signals based on deep learning approach," *Expert Syst. Appl.*, vol. 107, pp. 61–71, 2018, doi: 10.1016/j.eswa.2018.04.021.
2. A. T. Tzallas, M. G. Tsipouras, D. I. Fotiadis, and S. Member, "Epileptic seizure detection in EEGs using time—frequency analysis," *IEEE Trans. Inf. Technol. Biomed.*, vol. 13, no. 5, pp. 703–710, 2009.
3. D. Zhang and D. Zhang, "Wavelet transform," *Fundam. Image Data Min. Anal. Featur. Classif. Retr.*, pp. 35–44, 2019.
4. A. Anuragi and D. S. Sisodia, "Alcoholism detection using support vector machines and centered correntropy features of brain EEG signals," in *Proceedings of the International Conference on Inventive Computing and Informatics (ICICI)*, 2017, pp. 1021–1026.

5. A. Anuragi and D. S. Sisodia, "Alcohol use disorder detection using EEG Signal features and flexible analytical wavelet transform," *Biomed. Signal Process. Control*, vol. 52, pp. 384–393, 2018.

6. A. Anuragi and D. S. Sisodia, "Empirical wavelet transform based automated alcoholism detecting using EEG signal features," *Biomed. Signal Process. Control*, vol. 57, p. 101777, 2020.

7. V. Gupta, A. Bhattacharyya, and R. B. Pachori, "Classification of seizure and non-seizure EEG signals based on EMD-TQWT method," in *2017 22nd International Conference on Digital Signal Processing (DSP)*, August, 2017, pp. 1–5.

8. S. Saxena, C. Hemanth, and R. G. Sangeetha, "Classification of normal, seizure and seizure-free EEG signals using EMD and EWT," in *2017 International Conference On Nextgen Electronic Technologies: Silicon to Software, (ICNETS2)*, 2017, pp. 360–366, doi: 10.1109/ICNETS2.2017.8067961.

9. A. Bhattacharyya, R. Pachori, A. Upadhyay, and U. Acharya, "Tunable-Q wavelet transform based multiscale entropy measure for automated classification of epileptic EEG signals," *Appl. Sci.*, vol. 7, no. 4, p. 385, 2017.

10. A. I. Sharaf, M. A. El-Soud, and I. M. El-Henawy, "An automated approach for epilepsy detection based on tunable Q-Wavelet and firefly feature selection algorithm," *Int. J. Biomed. Imaging*, vol. 2018, no. 1, 2018, doi: 10.1155/2018/5812872.

11. U. Orhan, M. Hekim, and M. Ozer, "EEG signals classification using the K-means clustering and a multilayer perceptron neural network model," *Expert Syst. Appl.*, vol. 38, no. 10, pp. 13475–13481, 2011.

12. M. A. Hadj-Youcef, M. Adnane, and A. Bousbia-Salah, "Detection of epileptics during seizure free periods," *2013 8th Int. Work. Syst. Signal Process. Their Appl. (WoSSPA)*, 2013, pp. 209–213, doi: 10.1109/WoSSPA.2013.6602363.

13. A. Subasi, "EEG signal classification using wavelet feature extraction and a mixture of expert model," *Expert Syst. Appl.*, vol. 32, no. 4, pp. 1084–1093, 2007, doi: 10.1016/j.eswa.2006.02.005.

14. N. Nicolaou and J. Georgiou, "Detection of epileptic electroencephalogram based on permutation entropy and support vector machines," *Expert Syst. Appl.*, vol. 39, no. 1, pp. 202–209, 2012, doi: 10.1016/j.eswa.2011.07.008.

15. H. Akbari, S. S. Esmaili, and S. F. Zadeh, "Classification of seizure and seizure-free EEG signals based on empirical wavelet transform and phase space reconstruction," *arXiv Prepr. arXiv1903.09728*, 2019.

16. K. Jindal and R. Upadhyay, "Epileptic seizure detection from EEG signal using flexible analytical wavelet transform," in *Computer, Communications and Electronics (Comptelix), 2017 International Conference on*, 2017, pp. 67–72.

17. O. Singh and R. K. Sunkaria, "ECG signal denoising via empirical wavelet transform," *Australas. Phys. Eng. Sci. Med.*, vol. 40, no. 1, pp. 219–229, 2017, doi: 10.1007/s13246-016-0510-6.

18. R. G. Andrzejak, K. Lehnertz, F. Mormann, C. Rieke, P. David, and C. E. Elger, "Indications of nonlinear deterministic and finite-dimensional structures in time series of brain electrical activity: Dependence on recording region and brain state," *Phys. Rev. E—Stat. Physics, Plasmas, Fluids, Relat. Interdiscip. Top.*, vol. 64, no. 6, p. 8, 2001.

19. A. Anuragi, D. S. Sisodia, and R. B. Pachori, "Automated FBSE-EWT based learning framework for detection of epileptic seizures using time-segmented EEG signals," *Comput. Biol. Med.*, vol. 136, p. 104708, 2021.

20. J. Gilles, "Empirical wavelet transform," *IEEE Trans. signal Process.*, vol. 61, no. 16, pp. 3999–4010, 2013, doi: 10.1109/TSP.2013.2265222.

21. D. S. Sisodia, R. B. Pachori, and L. Garg, eds. *Handbook of research on advancements of artificial intelligence in healthcare engineering*. IGI Global, 2020. Doi: https://doi.org/10.4018/978-1-7998-2120-5

22. T. Chen and C. Guestrin, "XGBoost: A scalable tree boosting system," in *Proceedings of the 22nd ACM SIGKDD International Conference on Knowledge Discovery and Data Mining*, 2016, pp. 785–794. 2016. https://doi.org/10.1145/2939672.2939785.

23. G. Ke et al., "LightGBM: A highly efficient gradient boosting decision tree," in *NIPS'17: Proceedings of the 31st International Conference on Neural Information Processing Systems, Advances in Neural Information Processing Systems*, vol. 30, pp. 3149–3157, 2017.

24. A. V. Dorogush, V. Ershov, and A. Gulin, "CatBoost: Gradient boosting with categorical features support," *arXiv Prepr. arXiv1810.11363*, pp. 1–7, 2018.

25. R. J. Martis et al., "Application of empirical mode decomposition (EMD) for automated detection of epilepsy using EEG signals," *Int. J. Neural Syst.*, vol. 22, no. 6, pp. 1–16, 2012.

26. S. Patidar and T. Panigrahi, "Detection of epileptic seizure using Kraskov entropy applied on tunable-Q wavelet transform of EEG signals," *Biomed. Signal Process. Control*, vol. 34, pp. 74–80, 2017, doi: 10.1016/j.bspc.2017.01.001.

27. B. Biswas, S. Bhadra, M. K. Sanyal, and S. Das, *Epileptic seizure mining via novel empirical wavelet feature with J48 and KNN classifier*, vol. 695. Springer Singapore, 2018.

28. Y. Liu, W. Zhou, Q. Yuan, and S. Chen, "Automatic seizure detection using wavelet transform and SVM in long-term intracranial EEG," *IEEE Trans. Neural Syst. Rehabil. Eng.*, vol. 20, no. 6, pp. 749–755, 2012, doi: 10.1109/TNSRE.2012.2206054.

5 Automated Emotion Recognition from EEG Signals Using Machine Learning Algorithms in the Field of Ambient Assisted Living

Rohan Mandal, Uday Maji, and Saurabh Pal

5.1 INTRODUCTION

Human emotions are continuously on display whether or not we realize it, and in normal personal interactions, we are generally effective at recognizing mental states [1]. But in complicated situations, it can be very difficult to predict exact emotional states. For example, people might smile even when they are stressed.

Automated emotion recognition is a modern process of categorizing the human mental status through machine learning. It is actually inherent human intelligence to understand others' emotional states and communicate effectively, but there are many situations that may impede our correct determinations about others' emotional states, such as when people are suffering from undisclosed personal problems such as depression or anxiety [2]. Yet emotion analysis through human expression is crucial in varied areas such as autism detection, pain assessment, customer satisfaction analysis, and the education sector. Automated machine learning-based human emotion recognition can be performed in two ways: analyzing facial expressions and analyzing physiological signals such as EEG.

5.1.1 FACIAL-EXPRESSION-BASED EMOTION RECOGNITION

Facial-expression-based emotion detection has recently been developed with high accuracy and gained wide popularity. In this method, programs act like human brains and detect the emotions through image analysis based on different facial features (see Table 5.1) [3].

In fact, there has been a rising interest in enhancing every element of human–computer interaction in recent years. Affective sensing systems can learn useful information about the user's mental condition by analyzing their appearances. In addition to

DOI: 10.1201/9781003479970-5

TABLE 5.1

Facial Expressions of Different Emotions

Emotions	Facial expression	Image
Joy/Happiness	Corners of lips raised into a smile	
Amusement	Head: thrown back with jaw lifted. Eyes: crow's feet at the eyes indicating that the muscles have tightened. Mouth: open, jaw dropped with relaxed muscles	
Surprise	Dropped jaw, raised brows, wide eyes	
Anger	Lowered and burrowed eyebrows, intense gaze, raised chin	
Fear	Open mouth, wide eyes, furrowed brows	

TABLE 5.1 (*Continued*)
Facial Expressions of Different Emotions

Emotions	Facial expression	Image
Sadness	Furrowed brows, lip corners depressed	
Anxiety	Biting of the lips, facial movements involving elements of the fear expression, more eye blinks	
Disgust	Eyebrows pulled down, nose wrinkled, upper lip pulled up, lips loose	

emotions, other mental processes, social contact, and physiological signals are all reflected in facial expressions. A system to recognize human emotions is of utmost importance for creating emotional interactions between people and computers.

5.1.2 METHODOLOGIES FOR FACIAL-EXPRESSION-BASED EMOTION DETECTION

Automated facial expression recognition consists of four basic steps: face detection/ tracking, feature extraction, feature selection, and emotion classification, as shown in Figure 5.1. An essential stage in recognizing emotions is facial detection. It eliminates an image's irrelevant portions, and then it identifies key points known as facial landmarks (the Dlib library detects 68 facial landmarks). In images of the human face, the points of people's facial features are identified by their (x, y) coordinates, and these coordinates among prominent facial features like the eyes, brows, nose, mouth, and jawline can help in determining the emotion. For example, a surprised face shows greater distances between facial features than a calm face, so we can measure the Euclidian distance between the points, particularly on the mouth.

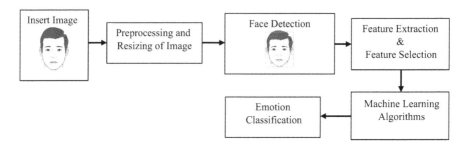

FIGURE 5.1 Block diagram of facial-expression-based emotion classification system.

Once the features from the input image are selected, a machine learning algorithm is used to categorize the emotion. The emotion-detection machine learning algorithms go through training to make sure the results are accurate and suitable. There are two approaches used to achieve accurate results: categorical and dimensional. Under a categorical method, a finite set of emotions fit into different set classes, such as contempt, sadness, shock, disgust, anger, happiness, and terror; in a dimensional method, the emotions exist on a spectrum and cannot be concretely defined. When using machine learning to model them, the approach option has some significant implications. A classifier would likely be developed and texts and images would be labelled with human emotions like "happy," "sad," etc. when a categorical model of human emotion is selected. However, outputs would have to be on a sliding scale if a dimensional model is selected. Once the training process is completed, images with unknown emotions are fed into the system, and the system judges the emotion based on the selected features.

The most extensive methods for determining feelings are based on geometrical facial features following the Facial Action Coding System. A facial-expression-based emotion recognition method [4] for assisted living utilizes the geometric features of the face to classify seven different emotions using a back propagating neural network. Alternate techniques like electroencephalography, electrocardiography, and skin impedance can be useful for obtaining accurate information to predict a person's mood, but they are very invasive because a patient must be attached to a device, in contrast to facial expression analysis, which is an entirely user-transparent process. It is therefore preferred that all the parts be as inexpensive as feasible because the finished system will be installed at home. There are additional noninvasive methods, like thermal imaging cameras, of gathering data for predicting a person's mood, but these tools are costlier than a regular webcam.

5.2 PHYSIOLOGICAL-SIGNAL-BASED EMOTION DETECTION

Numerous researchers suggested employing physiological signals such as skin temperature, electrodermal activity, photoplethysmogram (PPG), electroencephalography (EEG), galvanic skin response (GSR), blood volume pressure, and skin temperature to identify emotions [5, 6]. Studies have shown that a variety of indicators, including facial expressions, heart rate, body temperature,

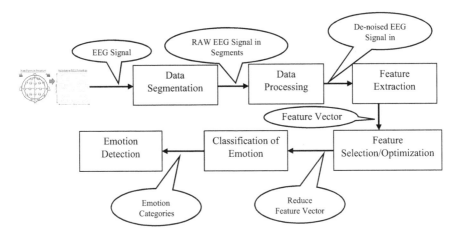

FIGURE 5.2   Block diagram of an EEG-based emotion detection system.

and electromyography, can be used to identify feelings. Signals from EEG have received a a great deal of attention due to their great efficiency and objectivity. Additionally, EEG is a quick and noninvasive technique that offers fairly accurate brain responses to internal or external stimuli, particularly for emotional stimuli [7]. A block diagram of EEG-based emotion detection technique is shown in Figure 5.2.

5.3   EEG DATA ACQUISITION

EEG is a graph that is created by well-calibrated electronic devices enhancing and recording the brain's spontaneous biological potential from the scalp. Through electrodes, which are often integrated in electrode caps, EEG captures the electrical activity of the brain [7]. The scalp's EEG data were captured using the 10–20 International System, a widely accepted technique for describing and applying where scalp electrodes should be placed during an EEG test, polysomnographic sleep study, and other related lab research. The 32 electrodes' placements are as follows: Fp1, AF3, F3, F7, FC5, FC1, C3, T7, CP5, CP1, P3, P7, PO3, O1, Oz, Pz, Fp2, AF4, Fz, F4, F8, FC6, FC2, Cz, C4, T8, CP6, CP2, P4, P8, PO4, and O2, shown in Figure 5.3. Here Fp represents front polar [4]; F, P, C, O, T represent frontal, parietal, central, occipital, and temporal, respectively. The odd numbers denote the left side of the brain, even numbers denote the right side, and z represents the middle portion of the brain.

5.3.1   The Characteristics of EEG

EEG is often divided into five categories based on frequency, and each group exhibits a different type of brain activity. Their frequency restrictions are categorized differently, as seen in Table 5.2.

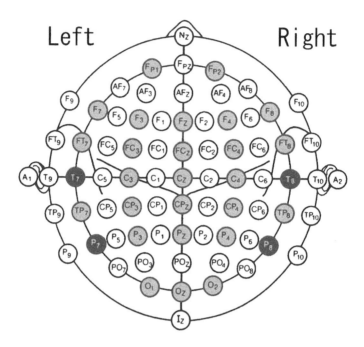

FIGURE 5.3   EEG electrode positions.

TABLE 5.2
Frequency Bands of EEG Signals

Frequency (Hz)	IFCN 1999 (II)	IPEG 2012	IFCN-2017 Glossary
Delta	0.5–4	1.5–<6	0.1–<4
Theta	5–7	6–<8.5	4–<8
Alpha	8–12	α1: 8.5–<10.5	8–13
		α2: 10.5–<12.5	
Beta	β1: 14–20	β1: 12.5–<18.5	14–30
	β2: 21–30	β2: 18.5–<21	
		β3: 21–<30	
Gamma	γ1: 30–40	30–<40	>30–80
	γ2: 40–. . .	γ1: 30–<65	
		γ2: 65–<90	
		γ3: 90–<135	

The EEG rhythms are connected to a variety of physiological and mental functions. Adults who are awake and resting, are in the alpha rhythm, particularly in the occipital region with bilateral synchrony, which is the primary rhythm of the brain at rest. At various stages of sleep, slower rhythms take the place of the alpha wave. While delta waves arise during deep sleep, theta waves appear during the early phases of sleep. In tense and worried subjects, high-frequency beta waves are present as background activity.

5.3.2 EEG Measuring Circuit

The measurement circuit will process the EEG signal after signal collection from electrodes. Instrumentation amplifiers, filters, A/D converters (ADC), signal processors, and power supplies make up the majority of the measurement circuit. Since the EEG signal is small and challenging for the ADC to detect, the amplifiers must have a high gain and common-mode rejection ratio in order to distinguish the EEG signal from background noise.

5.3.3 EEG Data Segmentation

While performing a medical test or scientific experiment, an EEG usually takes 20 to 40 minutes once the electrodes are in position. This is a significant amount of data that machine learning algorithms find extremely challenging to process. Therefore, before processing the signal, the effective segmentation of the obtained signal is always desired. Researchers have proposed different selected frames of suitable length of EEG signal for the processing.

5.3.4 Feature Extraction

EEG from the human brain is a complicated signal that lacks any discernible structure and is made up, as was previously said, of frequencies in a particular range. So researchers propose mainly frequency features like bandwidth, power spectral density, relative power, intensity weighted mean frequency, intensity weighted bandwidth, spectral edge frequency, spectral entropy, and peak frequency obtained from various transformation processes. Morphological features are rarely used for EEG signal analysis.

5.3.5 Feature Selection/Optimization

A predictive model's input variables are minimized through the process of feature selection. A machine learning algorithm's classification accuracy as well as execution speed are largely determined by the features made available by the system, so selecting useful features is crucial to the classification process. Therefore, prior to applying a final classification, it is always advised to utilize a proper feature selection and optimization technique. Many state-of-the-art feature selection/optimization techniques are used in EEG signal analysis, such as principal component analysis, linear regression, decision tree, support vector machine (SVM), K-means clustering, and hierarchical clustering.

5.3.6 Emotion Classification

Data is categorized into classes or categories using classification methods. Both structured and unstructured data can be used to conduct it. EEG-signal-based emotion classification mainly focuses on determining emotion valence (positive or negative) and specific emotions (happy, sad, anxiety, fear, etc.). With the aid of a system that has already been trained on a variety of emotions, unknown emotion is classified in this final stage. Classifier performance is dependent on the training data; hence it is always advised to properly train the classifier using reliable data.

5.4　RELATED WORK

Ambient assisted living (AAL) represents a special category of ambient intelligence. AAL enables the aged and especially able persons to live independently without human assistance in a smart home environment where different sensors and actuators are deployed to detect abnormality in human physical and mental health condition and generate corrective action accordingly. Many researchers have proposed activity-based AAL models. Anthony Fleury et al. [5] proposed an activity classification model where they used infrared, temperature and hygrometric sensors in a real flat to detect the presence and usage of some facilities by the person under study, a microphone for speech recognition, and a kinematic sensor for posture change and gait detection. Chernbumroong et al. used a temperature sensor and an altimeter with accelerometer [6], and Glascock and Kutzik used a heat-sensitive motion detector with contact and magnetic reed switches and vibration detectors [8].

Researchers have also used cameras to capture video or frames [9, 10] and even the IMU sensors of smartphones [11, 12] to detect activities of daily living. In human activity recognition (HAR) models following conventional machine learning approaches, the extracted and selected features from the sensor data are fed to classifiers like support vector machine [8, 11], k-nearest neighbor (KNN) [12], decision tree [13], artificial neural network [14], and many more. In contrast, in deep learning, the raw sensor data are directly fed to deep learning models like deep neural networks (DNNs) [6, 15] and long short-term memory (LSTM) [9] for HAR. Researchers have also used two-phase hybrid deep machine learning [16] using bi-directional LSTM and skip-chain conditional random field to recognize complex activity. However, both types of models found remarkable applications in the field of assisted living.

Nevertheless, human emotion recognition should be equally emphasized in AAL to determine the appropriate nature of assistance the person requires. Among the various available emotion detection techniques, EEG stands out because it represents electrical activity in the brain, the provenance of all human emotion. Classical machine learning algorithms are also applied for emotion detection from EEG signals. Authors of one such work [17] utilize different entropy-based features extracted from EEG signal to compare the performance of three classifiers: SVM, multilayer perceptron, and one-dimensional convolutional neural network, which outperformed the other two. Authors of another study [7] compared three different classifiers, quadratic discriminant analysis, KNN, and SVM, and found that SVM offered better accuracy in classifying four emotion classes. Asghar et al. [18] used ResNet-50, GoogLeNet, Inception V4

TABLE 5.3
Accuracy of Proposed Works by Researchers

Authors	Physiological signals used	Classification accuracy
Siddique et al.	EEG	99.57%
Oh et al. [20]	Facial expression, electrodermal signals	86.6%
Asghar et al. [18]	EEG	97.5%
Jeong et al. [21]	Facial expression	99.21% (using CK+ dataset)
Riaz et al. [22]	Facial expression	96.75% (using CK+ dataset)
Zenonos et al. [23]	ECG, PPG, skin temperature	70.6%
Ragot et al. [24]	ECG, electrodermal signals	70%
Zhang et al. [25]	Heart rate, blood volume pulse, skin temperature, and electrodermal activity	70%

and VGG-16 as DNNs for EEG feature extraction and feature clustering. Extreme learning machine wavelet auto encoder data augmentation [19] ensures high classification accuracy of different emotional states using ResNet18 deep learning architecture. A comparative study of different physiological signal based human emotion systems are shown in Table 5.3. From the table it is observed that emotion classification based on facial expression and EEG signal provides better result than any others. But EEG based classifiers are more reliable than facial expression as facial expression is disguisable whereas EEG is not. Therefore, it is worth saying that the best Emotion detection is possible by analysis of EEG signal other than any other physiological signal.

5.5 CHALLENGES IN EEG-BASED EMOTION DETECTION

Although EEG signals outperform any other human body vital signs in their usefulness for the detection of emotional states, there are several challenges with the analysis of EEG signals.

1. The standard EEG recording system comes with minimum 32 electrodes that need to be placed in the proper positions on the human scalp, and placement restricts mobility to some extent.
2. The large numbers of electrodes generate large amounts of data that need to be handled during classification.
3. EEG signals have very low amplitude, ranging between 20 to 100 μV, and quite high frequency band widths of 0.5 to 50Hz. This makes the signal very chaotic in nature.

Therefore, the recording and analysis of EEG signals are more complex than is interpreting other physiological signals like ECG or PPG that are normally periodic in nature. Against this background, we aimed with our study to reduce the necessary number of EEG recording electrodes by identifying selective zones of brain surface that are the most prominent for the detection of human emotional status.

5.6 A NOVEL APPROACH FOR EMOTION DETECTION FROM EEG SIGNALS

In this section, we discuss a novel approach for detecting two emotions, joy and anger. It is evident that negative emotions are harmful for health, and long-term negative emotions cause adverse effects on physical health, whereas positive emotional states promote good health. Therefore, discrimination between positive and negative emotions is very important in the field of AAL. We extracted the five significant frequency components from EEG signals using empirical mode decomposition (EMD) and fed this feature set to the classifier to distinguish between the two emotions. The two classifiers we used were SVM and bagging decision trees, and SVM secured 97.4% classification accuracy using the selected feature-set of our proposed model. Next we discuss our method.

5.6.1 DATA COLLECTION AND SEGMENTATION

We used a benchmark emotion dataset, MAHNOB-HCI [26], for this study. This database provides several physiological signals like EEG, ECG, galvanic skin response, respiration rate, and body temperature along with eye gaze data from video recordings of participants watching emotion-stimulating video clips from six different cameras for facial expression and head position detection. These recordings were made from 27 healthy participants while watching 20 different video clips. The participants' physiological responses to each excerpt were recorded and stored with 30 seconds of buffer before the start and after the end of the videos.

The EEG data was recorded using the Biosemi Active II system consisting of 32 active electrodes placed on participants scalp following the 10–20 International System as shown in Figure 5.3. The sampling rate of the EEG recording was 256 Hz. Participants identified their emotional states after each trial by completing a self-assessment report; the MAHNOB-HCI database contains a total of nine discrete emotion labels: anger, disgust, fear, sadness, anxiety, joy, surprise, amusement, and neutral. For this study, we used all the EEG data labeled as anger and joy in the database. As the excerpts vary in duration from approximately 30 seconds to 2 minutes plus the 30 seconds of padding at each start and end, each data file contains a huge amount of data; therefore, only a small segment of data needs to be identified for further analysis. Here, we used approximately the last 20 seconds of video run time assuming that the labeled emotional state would be most prominent during the ending of the clips.

5.6.2 Feature Extraction

Different researchers have used different time domain, frequency domain, and combined features of EEG signal for detection of emotional states. In this work, the selected segment of EEG signal was decomposed using EMD. Compared with the different forms of frequency analysis like fast-Fourier transform, discrete Fourier transform, and wavelet transform, EMD provides better information about the frequency components present in the signal, the size and amplitude of each frequency component, and other related information.

EMD [27] works on raw signals, decomposing them adaptively step by step. Each decomposed signal is formed by identifying and averaging all local extreme points and checking the termination condition based upon intrinsic mode functions (IMFs). If the condition is satisfied, the loop terminates; if not, the IMF is considered raw signal and again decomposed. This step repeats until the IMF condition is satisfied. Thus each step provides a decomposed signal for a different frequency component. While decomposing the EEG signal, EMD generates seven such components and selects five components that come under one of the significant EEG signal frequency bands, i.e., δ, θ, α, β, and γ. This was done for all 32 channels to form a $32 \times 5 = 160$-feature set from each EEG segment extracted from all available data for joy and anger. We tested the classification performance on the total feature set using two different classifiers: fine Gaussian SVM and bagging decision trees.

5.6.3 Feature Reduction

Here, we reduced the feature vector dimensions by selecting the features of selected EEG channels. Detailed study of the human central nervous system (CNS) is required for selecting the most informative EEG channels for emotion recognition. Then, the classifiers are used to automatically recognize different emotions using feature vectors extracted from the EEG signals [28, 29, 30].

5.6.4 Classification and Result

As we noted, we tested the performance of two classifiers: SVM using fine Gaussian kernel function and bagging decision tree, both very well-known classifier algorithms in the field of machine learning. We tested both classifiers for two cases: using the full 160-feature set and then a subset of 85 selected features. Their results and confusion matrices are shown in Figures 5.4 (a), 5.4 (b), 5.4 (c), and 5.4 (d).

The results indicate that the bagging decision tree gives similar classification accuracy and training times for the full and reduced feature sets, but it is more interesting to see that with SVM, both accuracy and training time significantly improved after feature reduction. So it is apt to say that the CNS-anatomy-based electrode selection technique can improve emotion detection accuracy while reducing the cost and complexity of EEG recording. There is ample room for more research in this area to identify more specific electrode positioning zones to effectively detect various emotional states.

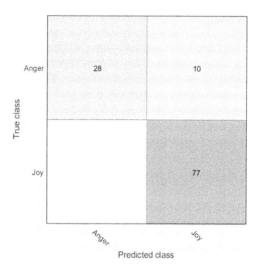

FIGURE 5.4A   Confusion matrix of fine Gaussian SVM classifier, accuracy 91.3%, 3-fold cross validation, training time 7.84 sec.

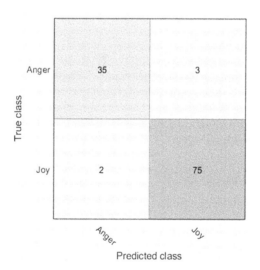

FIGURE 5.4B   Confusion matrix of bagging tree classifier, accuracy 95.7%, 3-fold cross validation, training time 3.1 sec.

5.7   CONCLUSION

EEG-signal-based ambient assisted living devices can be installed and used in a variety of intelligent surroundings. This technique can assist in identifying a person's mental health and providing the appropriate support. Not just in cases of physical disability but also in cases of poor mental health, humans require aid to

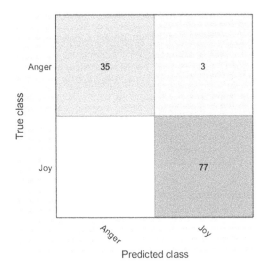

FIGURE 5.4C Confusion matrix of fine Gaussian SVM classifier, accuracy 97.4%, 3-fold cross validation, training time 1.25 sec.

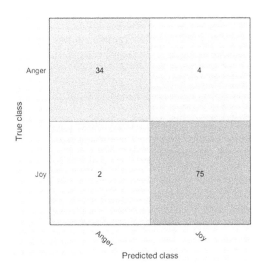

FIGURE 5.4D Confusion matrix of bagging tree classifier, accuracy 94.8%, 3-fold cross validation, training time 3.3 sec.

survive. In this work, we examined two major human emotions, joy and anger, and processed them with a minimal number of EEG leads. The results show that SVM classified the emotion with an accuracy of 97%. These findings support using EEG signals to recognize emotions with fewer leads and channels, which can improve quality of living.

REFERENCES

1. Fabiano D, Canavan SJ. Deformable Synthesis Model for Emotion Recognition. In Proceedings of the 2019 14th IEEE International Conference on Automatic Face & Gesture Recognition (FG 2019), Lille, France, 14–18 May 2019; pp. 1–5.
2. Zeng Z, Pantic M, Roisman G, Huang TS. A Survey of Affect Recognition Methods: Audio, Visual, and Spontaneous Expressions. IEEE Transactions on Pattern Analysis and Machine Intelligence 2008, 31, pp. 39–58. [CrossRef] [PubMed]
3. McDuff D, El Kaliouby R, Senechal T, Amr M, Cohn JF, Picard R. AMFED Facial Expression Dataset: Naturalistic and Spontaneous Facial Expressions Collected in–the–Wild. In Proceedings of the IEEE Conference on Computer Vision Pattern Recognition Workshops, Portland, OR, USA, 23–28 June 2013; pp. 881–888.
4. Xefteris S, Doulamis N, Andronikou V, Varvarigou T, Cambourakis G. Behavioral Biometrics in Assisted Living: A Methodology for Emotion Recognition. Engineering, Technology & Applied Science Research 2016, 6(4), pp. 1035–1044.
5. Fleury A, Vacher M, Noury N. SVM-Based Multi-Modal Classification of Activities of Daily Living in Health Smart Homes: Sensors, Algorithms and First Experimental Results. IEEE Transactions on Information Technology in Biomedicine, Institute of Electrical and Electronics Engineers 2010, 14 (2), pp. 274–283.
6. Chernbumroong S, Cang S, Atkins A, Yu H. Elderly Activities Recognition and Classification for Applications in Assisted Living. Expert Systems with Applications 2013, 40, pp. 1662–1674. Elsevier.
7. Ali M, Mosa AH, Al Machot F, Kyamakya K. EEG-based Emotion Recognition Approach for E-healthcare Applications. In 2016 Eighth International Conference on Ubiquitous and Future Networks (ICUFN) July 2016, pp. 946–950. IEEE.
8. Glascock AP, Kutzik DM. Behavioral Telemedicine: A New Approach to the Continuous Nonintrusive Monitoring of Activities of Daily Living. Telemedicine Journal 2000, 6 (1), pp. 33–44.
9. Guerra BMV, Schmid M, Beltrami G, Ramat S. Neural Networks for Automatic Posture Recognition in Ambient-Assisted Living. Sensors 2022, 22, p. 2609. https://doi.org/10.3390/s22072609
10. Liu Q, Chen E, Gao L, Liang C, Liu H. Energy-Guided Temporal Segmentation Network for Multimodal Human Action Recognition. Sensors 2020, 20(17), p. 4673.
11. Alemayoh TT, Lee JH, Okamoto S. New Sensor Data Structuring for Deeper Feature Extraction in Human Activity Recognition. Sensors 2021, 21, p. 2814. https://doi.org/10.3390/s21082814
12. Dirgová Luptáková I, Kubovčík M, Pospíchal J. Wearable Sensor-Based Human Activity Recognition with Transformer Model. Sensors 2022, 22, p. 1911. https://doi.org/10.3390/s22051911
13. Huang YC, Yi CW, Peng WC, Lin HC, Huang CY. A study on Multiple Wearable Sensors for Activity Recognition. In 2017 IEEE Conference on Dependable and Secure Computing, 2017; pp. 449–452.
14. Walse KH, Dharaskar RV, Thakare VM. PCA Based Optimal ANN Classifiers for Hu-man Activity Recognition Using Mobile Sensors Data. In First International Conference on Information and Communication Technology for Intelligent Systems: Volume 1, Smart Innovation, Systems and Technologies 50. https://doi.org/10.1007/978-3-319-30933-0_43
15. Jain A, Kanhangad V. Human Activity Classification in Smartphones Using Accel-Erometer and Gyroscope Sensors. IEEE Sensors Journal 2018, 18 (3), pp. 1169–1177.
16. Thapa K, Abdullah Al ZM, Lamichhane B, Yang SH. A Deep Machine Learning Method for Concurrent and Interleaved Human Activity Recognition. Sensors 2020, 20(20), p. 5770.

17. Mai ND, Lee BG, Chung WY. Affective Computing on Machine Learning-based Emotion Recognition using a Self-made EEG Device. Sensors 2021, 21(15), p. 5135.
18. Asghar MA, Khan MJ, Rizwan M, Mehmood RM, Kim SH. An Innovative Multi-model Neural Network Approach for Feature Selection in Emotion Recognition Using Deep Feature Clustering. Sensors 2020, 20(13), p. 3765.
19. Ari B, Siddique K, Alçin ÖF, Aslan M, Şengür A, Mehmood RM. Wavelet ELM-AE based Data Augmentation and Deep Learning for Efficient Emotion Recognition Using EEG Recordings. IEEE Access 2022, 10, pp. 72171–72181.
20. Oh G, Ryu J, Jeong E, Yang JH, Hwang S, Lee S, Lim S. DRER: Deep Learning–Based Driver's Real Emotion Recognizer. Sensors 2021, 21 (6), p. 2166. https://doi.org/10.3390/s21062166
21. Jeong D, Kim B-G, Dong S-Y. Deep Joint Spatiotemporal Network (DJSTN) for Efficient Facial Expression Recognition. Sensors 2020, 20 (7), p. 1936. https://doi.org/10.3390/s20071936
22. Riaz MN, Shen Y, Sohail M, Guo M. eXnet: An Efficient Approach for Emotion Recognition in the Wild. Sensors 2020, 20 (4), p. 1087. https://doi.org/10.3390/s20041087
23. Zenonos A, Khan A, Kalogridis G, Vatsikas S, Lewis T, Sooriyabandara M. Healthy Office: Mood Recognition at Work Using Smartphones and Wearable Sensors. In 2016 IEEE International Conference on Pervasive Computing and Communication Workshops (PerCom Workshops), Sydney, NSW, Australia, 2016; pp. 1–6. https://doi.org/10.1109/PERCOMW.2016.7457166.
24. Ragot M, Martin N, Em S, Pallamin N, Diverrez J-M. Emotion Recognition Using Physiological Signals: Laboratory vs. Wearable Sensors. Applied Human Factors and Ergonomics July 2017, pp. 813–822. Los Angeles, France. https://doi.org/10.1002/9781119183464. ⟨hal-01544007⟩
25. Zhang T, El Ali A, Wang C, Hanjalic A, Cesar P. CorrNet: Fine-Grained Emotion Recognition for Video Watching Using Wearable Physiological Sensors. Sensors 2021, 21 (1), p. 52. https://doi.org/10.3390/s21010052
26. Mohammad Soleymani, Jeroen Lichtenauer, Thierry Pun and Maja Pantic A Multimodal Database for Affect Recognition and Implicit Tagging. IEEE Transactions on Affective Computing 2011, 3 (1), pp. 42–55.
27. Shen Y, Wang P, Wang X, Sun K. Application of Empirical Mode Decomposition and Extreme Learning Machine Algorithms on Prediction of the Surface Vibration Signal. Energies 2021, 14 (22), p. 7519. https://doi.org/10.3390/en14227519
28. Shiripova LV, Myasnikov EV. Human Action Recognition Using Dimensionality Reduction and Support Vector Machine. In 5th International Conference on "Information Technology and Nanotechnology" (ITNT-2018), 2018.
29. Manosha Chathuramali KG, Rodrigo R. Faster Human Activity Recognition with SVM. In The International Conference on Advances in ICT for Emerging Regions—(ICTer 2012), 2012; pp. 197–203.
30. Yasin H, Hussain M, Weber A. Keys for Action: An Efficient Keyframe-Based Approach for 3D Action Recognition Using a Deep Neural Network, Sensors 2020, 20 (8), p. 2226. https://doi.org/10.3390/s20082226.

6 Automated Recognition of Human Emotions from EEG Signals Using Signal Processing and Machine Learning Techniques

Abdulhamit Subasi, Tuba Nur Subasi, and Oznur Ozaltin

6.1 INTRODUCTION

Automated emotion recognition entails recording facial expressions of intense feelings to identify sound indicators, eye gaze information, and physiological flags of the peripheral and central sensory systems to label and verify the feelings. Even if there are many techniques for emotion recognition, more research is still needed to improve acknowledgment accuracy and identify hidden areas of passionate behavior in the human mind. Multimodal data can be combined to enhance execution, lessen information grouping's vulnerability, or reduce uncertainty.

Understanding human emotion is essential in communication processes. Human–computer interaction (HCI) systems rely heavily on perception and the recognition of emotions. Several studies on emotion identification have been conducted. One of the main problems with emotion perception is how to explain human emotional states. To characterize emotional states, discrete models and dimensional models have been presented. While the dimensional model can be expanded, the discrete model often expresses the six fundamental emotional states of annoyance, disagreement, fright, happiness, surprise, and sadness (Zhang et al., 2016).

Many people have difficulty expressing their emotions because of differences in their cultures, social environments, ages, and life experiences. In various fields, including the field of medicine, the demand for sensors for recognizing human emotions has grown in order to treat a variety of neurological and psychological conditions and for keeping the patient's doctor informed. Because criminals occasionally manage to hide their fear and worry from the public, the security field also needs these sensors in criminal investigation cases. Without these sensors, the security team is more vulnerable to attack.

In response to increased need, we aim to develop the most significant technologies for recording human emotions through our research. Understanding and evaluating

DOI: 10.1201/9781003479970-6

human emotions is essential when analyzing human behavior since emotions are the most significant factors that influence a person's behaviors. For the field of affective computing to advance, an understanding of human emotions and their effects is essential. HCI requires emotional intelligence, knowledge of human emotional experience and the connection between emotional experience and affective expression. Human emotions are influenced by a variety of activities, like viewing movies and listening to music. If we could use sensor technology to capture these emotional shifts when people are exposed to different stimuli, it would help us learn more about how consumers make emotional decisions (Daimi & Saha, 2014).

Brain signals in reaction to audio music recordings have been studied for human emotion identification. Using audio music gives those who are visually impaired the ability to recognize emotions, which is a definite benefit. Researchers obtained 13 characteristics from three separate domains, and using four different classifier types, divided them into four different emotions. Bhatti et al. (2016) showed that multilayer perceptron (MLP) provided higher accuracy than support vector machine (SVM) and K-nearest neighbors (K-NN) classifiers, and that happy and sad were the most straightforward emotions to categorize regardless of the classifiers employed. On the other hand, love and anger emotions are challenging to identify.

The features extracted from EEG typically change dramatically, whereas emotion states fluctuate steadily, as EEG is an unstable voltage signal. The labels of emotion samples can only be predicted by existing studies; they cannot capture the pattern of changing emotions. Wang et al. (2014) analyzed three different EEG emotion-specific feature types and evaluated how well two emotion states could be classified from the EEG data of six people in order to confirm the efficacy of the suggested techniques. They provided a brief summary of related research on emotion models, the role of movies in inducing emotions, and various EEG-based emotion classification techniques.

Because of the fuzzy boundaries, emotion recognition from EEG is highly difficult. Zheng et al. (2017) provided a brief summary of relevant research on EEG-based emotion detection as well as the discoveries of consistent patterns for different emotions. To identify persistent patterns and brain signatures for different emotions, as well as to assess the consistency of their emotion identification models across time, they employed time-frequency analysis.

Collaboration between humans and machines occurs in a variety of settings, including in the workplace and in health. In order to accomplish high levels of user happiness, researchers in the field of work environments and smart systems are continually working to increase the effectiveness and flexibility of the interaction between computers and humans. In order to fully grasp human communication, HCI systems must include the capacity to adapt to computers. Humans can convey emotions verbally or nonverbally, so HCI systems assist in comprehending and analyzing nonverbal human thought, as well as in understanding emotional responses of an individual (Yin et al., 2017).

Some researchers concur that data derived from emotion sensors has to be improved because it is erroneous. Because some sensations, such as fear and anxiety, closely resemble other feelings, it can be challenging to distinguish them. It is necessary to develop EEG to swiftly identify human emotions and provide results that are more accurate. Because we want to increase the precision of extracted emotions to be more dependable in various domains, it is crucial to understand the most recent technologies for data extraction and representation. We must utilize all new emotion sensors in light of the rapidly advancing technology if we want to deliver accurate data. In this chapter, we compare the performance of different ensemble classifiers for emotional EEG signal classification. Due to the poor signal-to-noise ratio and substantial subject variability of EEG data, hand-designed features generally underperform. Ensemble classifiers can capture the intrinsic characteristics of EEG signals. The proposed model incorporates DWT feature extraction and ensemble classification.

6.2 LITERATURE REVIEW

A person's emotional experiences influence the multimedia content they choose. Daimi and Saha (2014) employed EEG to monitor brain activity and find emotions in various emotional circumstances because brain electrical activity carries the emotional signals required for emotion recognition. First, they extracted time–frequency emotion characteristics using dual-tree complex wavelet transform (DT-CWT). Then, utilizing leave-one-out cross-validation, they chose the non-recurring emotional traits via singular value decomposition. The outcomes demonstrated that EEG signals correlated with participants' subjective assessments of emotional traits.

Zheng et al. (2014) suggested a recognition system for emotions based on pattern identification over time from EEG. They employed machine learning to assess stable EEG patterns over time. Numerous activation patterns have been linked to different emotions, although it is unclear how stable these patterns are over time. Zheng et al. aimed to pinpoint EEG stability in emotion detection. With the DEAP dataset and a recently created dataset, they systematically evaluated numerous common extractions of feature, selection of feature, smoothing of feature, and pattern classification and found that consistent patterns indicate stability over time. The effectiveness of the suggested system for recognizing emotions demonstrates that brain patterns are largely consistent both within and across sessions.

Iacoviello et al. (2015) proposed a method for classifying EEG signals that would result from self-induced emotions in a way that is generalized, effective, parametric, and fully automatic in real time. The idea took into account specific traits of low-amplitude signals of self-induced emotions, which account for 15% of an actual emotion. The authors employed wavelet transform, principal component analysis (PCA), and SVM for analysis and categorization using a multi-emotion brain–computer interface (BCI). This method is characterized as a machine learning framework and is a two-stage algorithm with parameters that specifically focus on multi-class categorization. Calibration, the initial step, is an offline procedure that establishes the features, signal processing, and classifier training. Testing of data is part of the real-time second stage. Redundancy is avoided using PCA, and SVM is used to classify the features. The outcomes of this suggested strategy have a high proportion of over 90%, which is quite encouraging.

According to Othman et al. (2013), the field of BCIs has produced a number of examples showing how EEG emotion identification algorithms can be used to compute human emotion. These studies, however, have mostly concentrated on identifying a distinct emotion model for each separable basic emotion that is well-known among psychologists. There is currently no consensus on the number of fundamental emotions, although the majority of experts agree to accept six: sadness, anger, happiness, fear, disgust, and surprise. The distinct emotion model has received criticism for failing to capture how real emotions interact with each other. But due to its ease of usage, it was embraced.

An important component of diagnosing patients' mental health using EEG is the ability to recognize emotions, but EEG results are unreliable, and as a result, classification performance suffers. Methods for domain adaptation offer a practical means of reducing the disparity in the marginal distribution. Adaptive subspace feature matching is the name of the innovative method suggested by Chai et al. (2017). It reduces domain discrepancy and performance deterioration by combining the marginal and conditional distributions within a merged framework.

Verma and Tiwary (2014) proposed a framework for emotion classification and recognition using physiological signals. They used EEGs with 32 channels and peripheral signals (8 channels) from the DEAP database, including the electromyogram, electrooculogram, blood volume pressure, respiration pattern, and skin temperature to evaluate multimodal physiological signals. A continuous three-dimensional representation model for emotions was also proposed by the researchers after discussing theories of emotion modeling based on (i) fundamental emotions, (ii) cognitive evaluation and physiological response, and (iii) the dimensional method. The suggested model was validated using a clustering experiment on the given valence, arousal, and dominance values of different emotions. The SVM, MLP, KNN, and MMC classifiers' average accuracies from this framework were 81.45%, 74.37%, 57.74%, and 75.94%, respectively. 'Depressing' has the highest accuracy with 85.46% when using SVM. The outcomes of this framework have demonstrated a high level of 85% accuracy with 32 people and 13 emotions, demonstrating the promise of the multimodal fusion technique.

Lahane and Sangaiah (2015) proposed EEG approaches to address the growing trend of using computers to capture emotions. Their study relies on EEG signal analysis to determine emotions from brain activity in humans. The scalp of the brain is used to record EEG signals, which are then evaluated in reaction to a number of the four primary emotions listed on the IAPS emotion motivations. To determine the emotional state of the test participant, features from the EEG signals are extracted utilizing kernel density estimation (KDE) and ranked by an artificial neural network (ANN) classifier. Results are gathered to demonstrate that the suggested improved KDE offers more accurate results.

The EEG recording of emotions has reportedly become one of the most difficult tasks, according to the researchers. By identifying the phase correlations between frequency elements and describing the non-Gaussian data included in the EEG signals, Kumar et al. (2016) suggested a method for obtaining phase data using bispectral analysis. High scores were obtained to classify sensations of arousal in this study's valence–arousal emotion model, which the researchers used to quantify the number of emotions. Some emotion sensors may not accurately register all emotions. Therefore, it is crucial to build the models that are employed.

Xu et al. (2018) studied theories for modeling emotions as well as methods for measuring emotions. They examined developing BCI for emotion identification. The induction techniques for learning emotions and associated problems are also covered. Challenges for further learning emotion research are presented at the survey's conclusion. Hemanth and Anitha (2018) overcame the shortcomings of conventional neural networks in terms of computational complexity and accuracy. Circular back propagation neural network and deep Kohonen neural network were proposed to investigate EEG data to categorize various human moods. They confirmed suggestions' superior performance over the related methodologies.

Chakladar and Chakraborty (2018) introduced the technique for dimension reduction known as correlation-based subset selection. The categorization process was then carried out utilizing the higher-order statistics attributes of the smaller set of channels. They divided emotions into four categories: positive, negative, furious, and harmonious. Their suggested algorithm runs in $O(n2 + 2n)$ time. With the smaller set of channels, this model's classification accuracy was 82%. Finally, they compare their suggested model with a few well-known emotion categorization methods, and the outcome demonstrates that their model performs significantly better than all the earlier models. However, the suggested paradigm enables people with physical disabilities to effectively and quickly convey their emotions.

A spatial filtering technique based on independent component analysis (ICA) was proposed by Su et al. (2018). Specifically, they created a stimulus selection approach based on the normalized valence/arousal space model in order to locate effective video stimuli to elicit emotions. The optimal component-to-electrode mapping patterns in various emotional states were then utilized to develop an ICA-based spatial filter. To extract feature parameters, the devised filter linearly projected EEG data with five emotional intensities in terms of arousal and valence dimensions. Finally, emotions were categorized using SVM. The results of the experiment validated the ICA-based spatial filtering method proposed for identifying emotional intensity.

Subramanian et al. (2018) examined the linear and nonlinear physiological correlates of emotion and personality, correlations between users' emotional assessments and personality scales in the context of earlier observations. According to the analysis, nonlinear statistics rather than linear ones are a better fit for describing the link between emotions and personality. Finally, they used physiological markers to recognize binary emotions and personality traits. Overall, the results of the experiments support the idea that personality variations can be more clearly seen when comparing user reactions to emotionally homogeneous videos. Above-chance detection is also attained for both affective and personality characteristics.

Katsigiannis and Ramzan (2018) presented a multimodal database (DREAMER) consisting of electrocardiogram and EEG signals acquired during affect elicitation by means of audiovisual stimuli. Following each stimulus, 23 participants reported their signals as well as their assessments of their own efficient states in terms of valence, dominance, and arousal. Affective computing techniques may be used in commonplace applications thanks to portable, wearable, wireless, affordable, off-the-shelf technology that was used to record all of the data. These findings point to the potential of low-cost devices for applications involving affect identification.

Yang et al. (2018) proposed a hierarchical network topology with subnetwork nodes to recognize three human emotions, positive, neutral, and negative. Each embedded subnetwork node, which is made up of hundreds of hidden nodes, can operate as a separate hidden layer for feature representation. Other cutting-edge methods are contrasted with the suggested method. The experimental results from two separate EEG datasets indicate that employing the suggested strategy with both single and multiple modalities yields a positive result.

Zhao et al. (2018) identified a person's personality traits by examining their brain activity while they were exposed to emotionally charged content. The study involved 37 participants who were shown seven standardized film segments portraying real-life emotional events, with a focus on seven distinct emotions. The findings indicate that the model exhibits superior classification efficacy for extraversion (81.08%), agreeability (86.11%), and conscientiousness (80.56%). Additionally, the model demonstrates greater classification accuracy for neuroticism (78.38–81.08%) when negative emotions, excluding disgust, are elicited in comparison to positive emotions. Furthermore, the model attains the highest classification accuracy for openness (83.78%) when a repulsive film clip is exhibited. The results of this study demonstrate that the accuracy of personality classification can be enhanced by utilizing EEG signals, compared with the use of advanced explicit behavioral indicators.

Liu et al. (2018) generated a database with 16 emotive movie clips chosen from more than a thousand film snippets. They proposed a system that can detect a person's emotional states in real time by analyzing their brain waves in response to the film clips and used it to investigate 30 participants who viewed the 16 clips. The clips had been designed to portray a range of seven distinct emotions, as well as neutrality and real-life emotional events. The results demonstrated the superiority of utilizing EEG signals for immediate recognition of emotions compared with current cutting-edge algorithms, as evidenced by higher classification accuracy and the ability to distinguish between closely related discrete emotions.

Subasi et al. (2021) introduced an innovative framework for automated emotion recognition that employs EEG signals. The method comprised four stages, namely, reprocessing, feature extraction, dimension reduction, and classification. During the preprocessing phase, multiscale principal component analysis (MSPCA) was employed using a Symlets-4 filter for the purpose of noise reduction. The authors' feature extraction process involved utilizing a tunable Q wavelet transform (TQWT). They reduced dimensions with six distinct statistical techniques and employed the rotation forest ensemble (RFE) classifier during the classification stage. The proposed framework attained a classification accuracy of more than 93% through the utilization of RFE + SVM. The findings unambiguously demonstrate that the emotion recognition framework utilizing TQWT and RFE, as proposed, constitutes a proficient method for discerning emotions through EEG signals.

Tuncer et al. (2021) developed a novel method for identifying emotions by employing EEG signals. The primary objective of this method is to offer a precise framework for emotion recognition that integrates both manual feature engineering and a deep classifier. The architecture employs a network that generates fused features at

multiple levels. The three primary stages of this network are TQWT, statistical feature extraction, and nonlinear textural feature extraction. They employed TQWT to
perform signal decomposition on the EEG data, resulting in the creation of multiple
sub-bands. This, in turn, facilitates the development of a feature-generating network
with multiple levels.

The LED block cipher employed an S-box in its nonlinear feature generation
process, which produced a distinct Led pattern. Furthermore, they extracted statistical features using widely recognized statistical moments. The ReliefF and
iterative Chi2 methods are employed for the purpose of choosing the most informative features. The proposed model was developed utilizing the GAMEEMO and
DREAMER datasets, two EEG emotion datasets. Their proposed manual learning
network achieved classification accuracies of 94.58%, 92.86%, and 94.44% for the
arousal, dominance, and valence cases of the DREAMER dataset, respectively.
Additionally, the proposed model exhibits a classification accuracy of 99.29% for
the GAMEEMO dataset.

Emotions have been recognized as the primary factor in establishing meaningful relationships between people and machines. The need for accurately identifying
emotions has greatly expanded as a result of EEG breakthroughs, particularly in
the use of portable and affordable wearable EEG devices. However, there is still
little scientific research and expertise available on EEG-based emotion recognition.
Hancer and Subasi (2022) proposed an EEG-based framework for emotion recognition to address this problem. Preprocessing, feature selection, feature extraction, and
classification are the processes that make up the suggested framework. MSPCA and
the Sysmlets-4 filter are employed during the preprocessing stage. For the feature
extraction stage, they used DTCWT. Several statistical criteria are used to lower the
feature dimension size. Due to their promising performance in classification challenges, Hancer and Subasi chose ensemble classifiers for the last stage. The proposed
system used a random subspace ensemble classifier to attain accuracy of almost
96.8%. The conclusion that follows is that the proposed EEG-based paradigm does a
good job of recognizing emotions.

6.3   MATERIALS AND METHODS

6.3.1   THE DATASET

For this study, we used the SJTU Emotion EEG Dataset (SEED), which is available
to the public. While 15 volunteers watched emotional movie clips while wearing
an EEG apparatus designed to measure emotion, the signals from their EEGs were
recorded. The participants were instructed to write down their feelings about each
film clip on a questionnaire after seeing it to provide feedback. To explore neuronal signatures and consistent patterns across sessions and individuals, each subject
needed to complete the experiments for all three sessions; there was a week or more
between sessions. EEG data and facial videos were recorded simultaneously. The
worldwide 10–20 system was utilized to record EEG using a 62-channel active AgCl
electrode cap and an ESI NeuroScan system with an average sampling frequency of
1000 Hz (Zheng & Lu, 2015).

6.3.2 Signal Denoising with MSPCA

Wavelet analysis's ability to extract deterministic features and PCA's ability to remove nearly all relationships between auto-correlated measurements are combined in MSPCA, which removes relationships between variables by extracting linear relationships. By integrating the outcomes at pertinent scales, MSPCA determines the PCA of the wavelet coefficients at each scale. MSPCA is well-suited for generating data from events that exhibit temporal and frequency variations due to its inherent characteristics.

The MSPCA process monitoring is connected to control the scales at which noteworthy events are discovered. In MSPCA, process monitoring is a counterpoint to filtering the scores and residuals and fitting the detection thresholds for the quickest identification of predictable changes in the data. MSPCA is a suitable method for monitoring auto-correlated observations, even without the need for matrix augmentation or time-series modeling, due to the near de-correlation of wavelet coefficients. Furthermore, the utilization of MSPCA checking enhances the capacity to identify deterministic alterations while concurrently detecting features indicative of anomalous operation (Bakshi, 1998).

The optimization of various techniques is contingent upon the selection of pertinent coefficients and the representation of input and output variables in the wavelet domain. Multiscale modeling is a form of error-in-variables modeling due to its simultaneous consideration of error elimination and model parameter selection. In MSPCA, the process of wavelet decomposition effectively eliminates the correlation between the stochastic measurements, while PCA serves to decorrelate the relationship between the variables. Employing distinct PCA models for the coefficients at each scale can enhance the efficacy of noise elimination by utilizing scale-dependent threshold values, particularly in cases where the noise is non-white or scale-dependent.

At each scale, thresholding criteria are utilized based on the specific application to select wavelet coefficients and a particular subset of principal components. As an illustration, in the context of MSPCA, the objective of accomplishing multivariate noise removal may entail the utilization of univariate thresholding techniques to ascertain the threshold of the latent variables at each scale. The selection of the principal components is contingent upon the characteristics of the interrelations among the variables. The components with the same number at each scale are chosen so that the wavelet decomposition won't have an impact on the relationship between the variables.

In MSPCA, the results obtained from improving a non-dependent PCA model at each scale are aggregated as a final step. The process involves the reconstruction of a signal that is contingent upon the chosen coefficients at each scale, followed by PCA on the reconstructed signal, and the subsequent removal of any extraneous components. This particular procedure enhances the efficacy of noise elimination while preserving the orthonormality of the scores and loadings pertaining to the reconstructed signal. The process of selecting wavelet coefficients with the highest relevance involves thresholding input and output latent variables. By adding coefficients beginning at the most crucial stopping point, which cross-validation identifies, this model may be enhanced. It is observed that when this approach is applied to real-world issues, it may produce more precise parameters, higher prediction abilities, and models that offer more physical understanding of the system being modeled (Bakshi, 1998; Alsberg et al., 1998; Trygg & Wold, 1998). Multiscale modeling lacks a robust statistical foundation, and the thresholding and modeling phases are not properly merged. The present approaches can be replaced by multiscale modeling that thresholds the latent variables. The following optimization problem must be solved in order to effectively combine modeling with thresholding (Bakshi, 1999).

6.3.3 FEATURE EXTRACTION WITH DWT

Wavelet-based methods for EEG analysis are deemed suitable for examining diverse nonstationary signals. DWT is utilized to scale and shift the mother wavelet, thereby separating a discrete-time signal $x[k]$ into a collection of signals or wavelet coefficients. To initiate the decomposition process of the discrete wavelet transform, it is imperative to select the appropriate quantity of wavelet decomposition or scale levels, denoted as j_m. At the primary scale level, the value of j is 1, and the signal $x[k]$ is concurrently transmitted to both the high-pass filter ($h[\cdot]$) and the low-pass filter ($l[\cdot]$). Subsequently, the down-sampling process (2) follows. The exhibition of the variable j for the output of every level is demonstrated in the form of two signals, namely Detail (D_j) and Approximation (A_j):

$$D_j[i] = \sum_k x[k] \cdot h[2 \cdot i - k] \tag{6.1}$$

$$A_j[i] = \sum_k x[k] \cdot l[2 \cdot i - k] \tag{6.2}$$

where 1 raises j and the approximation A_j is displayed in order to advance to the subsequent level. The above-described generation process of D_j and A_j is repeated until j does not exceed j_m (Ghorbanian et al., 2012).

6.3.4 DIMENSION REDUCTION

Using first-, second-, third-, and fourth-order statistics of the sub-bands is one method of reducing dimension; the reduced feature set is calculated from the sub-bands of the wavelet decomposition. The following six statistical characteristics are implemented for EEG signal classification: absolute mean, average power, standard deviation, ratio of the absolute means of adjacent sub-bands, skewness, and kurtosis of the signal coefficients in every sub-band.

6.3.5 MODEL ENSEMBLES

Ensemble modeling aims to develop and integrate multiple inductive models for a single domain, with the potential to improve the accuracy of prediction for the majority or all of them. For this progress to be possible, particular models' strengths should be maintained or enhanced, while their deficiencies should be eliminated or minimized. It turns out that several hundred or so models, even those with poor performance, can collaborate to produce accurate forecasts.

The two primary predictive modeling tasks, classification and regression, are both amenable to ensemble modeling. Even if each individual model is perfectly legible by humans, it may be possible to achieve a significant improvement over single models by investing significantly more computing time in the generation of multiple models and sacrificing overall human readability. The most popular strategies for base model development and aggregation will be covered in this chapter in order to make use of this potential for improved predictive capability. The majority of this material applies to both the classification and regression tasks because the former is primarily task independent, and the latter's task-specific elements are sufficiently simple and isolated. Model ensembles are, however, more frequently and effectively employed in the former case, and certain ensemble modeling strategies created especially for classification will also be explored (Cichosz, 2014).

6.3.5.1 AdaBoost

AdaBoost (for adaptive boosting) was one of the first boosting techniques recommended by Freund and Schapire (1995). The most well-known program of this type is still the algorithmic one. The procedure for creating a collection of models is based on the training dataset and the probabilities Px assigned to each item x in the training set. The completion of the learning phase can vary depending on the chosen approach, either weighting or sampling, which may be more or less popular.

The learning process is provided with the training dataset and associated weights, on the condition that it acknowledges the assigned weights for each object within the training data. Subsequently, a data sample is extracted from the training set while accounting for the probability distribution p and subsequently transmitted to the learning machine. Given that the original distribution is uniform, the bagging method may be used to construct D1. For each new model MI created after that, the probability distribution is changed and the error is calculated such that future learning procedures focus more on the wrongly classified data items than the successfully recognized ones. The idea of distribution changes is to multiply by $\beta i < 1$ (decrease) the chances of correctly classified objects in the simplest way possible such that the sum of incorrectly classified object probabilities is equal to 0.5, which is the probability of correctly classified object probabilities. The ensemble's ultimate judgment is formed from the sum of the judgments of its individual members, with bigger weights for extra right models and less weights for subpar classifiers. The weights listed in achieve the desired result.

The initial formulation created specifically for two-class situations is modified in M1. One of AdaBoost's biggest weaknesses in multiclass classification is its requirement that strong learners' faults be taken into account. Otherwise, the variables i would increase beyond one and the purpose of boosting would be reversed, decreasing rather than increasing the likelihood of objects being misclassified. Freund and Schapire (1995) suggested an algorithmic rule called AdaBoost to deal with this. M2 uses a different type of model (fuzzy classification, which assigns a value between 0 and 1 to each class rather than crisply assigning them), a different method for handling the probability distribution, and a modified ensemble decision function. Freund and Schapire discuss the process in more detail, but in brief, the AdaBoost algorithm is terminated, and the ensemble is made up of the models found up to this point without the final one in all significant error after the error of the succeeding model exceeds 0.5. In such cases, some writers advise resuming the probabilities (returning to the uniform distribution) rather than stopping the operation or applying the standard probability modification procedure.

Following the probability distribution's reset, the following sample is once again a bootstrap sample. Each sample may be a bootstrap sample, and the member models will be the same as in bagging if each model has an error larger than 0.5. The final model's weighted decision function, as opposed to the majority vote employed in bagging, is the only distinction between this situation and bagging. AdaBoost cannot proceed to the next stage if a model has completely classified the training data ($\varepsilon i = 0$), as the distribution has degenerated. Additionally, under these circumstances, some writers advise resetting the probability distribution and creating the following model using a new bootstrap sample (Grąbczewski, 2014).

Given bagging and AdaBoost appear to operate in different ways, have distinct effects, and both have the strongest effects on the first few committee members, it is possible to benefit from combining them. The fact that the mechanisms work in diverse ways means that they might work better together than they would separately. Bagging primarily lowers variance, whereas AdaBoost also lowers bias, and there is evidence that bagging is more efficient at lowering variance than AdaBoost (Bauer & Kohavi, 1999), so their combination may be able to maintain AdaBoost's bias reduction while boosting the variance reduction that AdaBoost has already attained.

It is conceivable that even extremely small committees created by combining the methods may succeed, as the first few members of each committee receive the majority of the benefits of each methodology. Simply forming two subcommittees, one with members recruited through boosting and the other with members recruited through bagging, would allow the two to be joined. But how to aggregate the votes from the two subcommittees raises a challenging problem. Because AdaBoost gives its committee members' votes weight while bagging does not, the votes of the members of each committee cannot be compared.

For this study, we looked into the creation of several subcommittees, each of which employs AdaBoost, rather than pursuing this technique and dealing with this problem. Instead of bagging per se, which would have reduced the amount of training cases available to form each subcommittee, we utilized wagging (Bauer & Kohavi, 1999). Maintaining all instances in the training set was important because, according to Quinlan (1996), AdaBoost using reweighting benefits from access to all training examples throughout the induction of each committee member. Finally, AdaBoost3 can be seen as establishing wagging committees, which are a subset of bagging committees. The number and size of subcommittees that should be formed for a single run must be decided.

6.3.5.2 MultiBoost

The current MultiBoosting implementation, MultiBoost, accepts a single committee size T as an argument, from which it sets the quantity of subcommittees and their sizes to T by default. This is carried out in the absence of any a priori justification for selecting specific values for these parameters. Since both of these values must be whole numbers, the result must be rounded off. This is done by creating a target final subcommittee member index, where each final committee member is given an index, starting at 1, to make implementation easier. As a result, the size of the succeeding subcommittee can be increased in the event that a subcommittee is prematurely terminated as a result of an abnormally high or low error. In the case that the preceding subcommittee is prematurely dissolved, the new subcommittee's objective is to fulfill the full complement of committee members.

The process is repeated with the addition of further subcommittees until the required overall committee size is reached, and if the additional subcommittee fails to accomplish its objective either. The ability to train subcommittees concurrently gives MultiBoost a possible computational advantage over AdaBoost, although doing so would call for a modification in how subcommittee learning is handled when it is terminated too soon. The bias and variance reduction capabilities of each of the committee learning algorithms that make up this method may be inherited. The potential for parallel processing is constrained by the AdaBoost algorithm' inherent sequential nature. MultiBoost inherits from each classifier learned with wagging at the subcommittee level its independence, enabling parallel computation (Webb, 2000).

6.3.5.3   The Random Subspace Method

Ho (1998) suggested the combining method known as the random subspace method (RSM). One can alter the training data in the RSM as well. This alteration is made in the feature space, though. In the training sample set $X = (X_1, X_2, \ldots, X_n)$, each training item X_i $(i = 1, \ldots, n)$ should be a p-dimensional vector $X_i = (x_{i1}, x_{i2}, \ldots, x_{ip})$, defined by p features (components). One chooses $r < p$ features at random from the p-dimensional data collection X when using the RSM. Thus, one obtains the original p-dimensional feature space's r-dimensional random subspace. Consequently, the r-dimensional training samples $X=[X_1; \ldots; X_n]$ make up the modified training set, $\widetilde{X^b} = \widetilde{X_1^b}, \widetilde{X_2^b} \ldots, \widetilde{X_n^b}$. Then, in the final decision rule, one may create classifiers in the random subspaces and merge them by simple majority vote. As a result, the RSM is structured as follows: $X=[x_1; \ldots; x_n]$
   1. Iterate till $b = 1, 2, \ldots, B$:

   (a) Choose a subspace $\widetilde{X^b}$ of dimensionality r from the initial feature space X, where X is of dimensionality p and the subspace is chosen randomly.

   (b) Develop a classifier denoted as $C^b(x)$ that operates within the domain $\widetilde{X^b}$ and possesses a decision boundary of $C^b(x) = 0$.

   2. Using simple majority voting, combine classifiers $C^b(x)$, $b = 1, 2, \ldots, B$, into a final decision rule.

The RSM may derive satisfaction from utilizing randomized subspaces for each building and consolidating the classifiers. When the number of training instances is significantly smaller than the feature space, implementing classifiers in random subspaces can effectively address the issue of small sample size. The spatial property of the topological space is smaller than that of the original feature space, while the number of training objects remains constant. Consequently, the size of the coaching sample will be relatively large. When there are multiple redundant options in knowledge, it is possible to achieve higher classifiers in random subspaces than in the original feature area. The aggregated classification approach may outperform a single classifier trained on the initial training set across the entire feature space (Skurichina & Duin, 2002).

6.4 RESULTS

The primary objective of EEG signal processing in BCI is to implement a framework for creating contactless interaction between a human and an external device, such as a speller. This infrastructure can aid disabled individuals and enable them to engage in their regular activities. The block diagram of such a BCI architecture is depicted in Figure 6.1. Regarding this figure, MSPCA is used to eradicate various types of EEG sounds and artifacts following signal acquisition. In the second stage, the pertinent features are extracted to train the classifier using various time-frequency-based methods. Then, dimension reduction is used to eliminate unnecessary features, resulting in improved classification accuracy. When the classification task is finished, the associated commands to control a device are formed (Ghaemi et al., 2017).

6.4.1 Discussion

EEG signals have been shown in a great number of studies to be able to give helpful information on activities occurring in the brain in order to identify emotional states (Zhang et al., 2016; Pham et al., 2015; Petrantonakis & Hadjileontiadis, 2011; Bajaj &

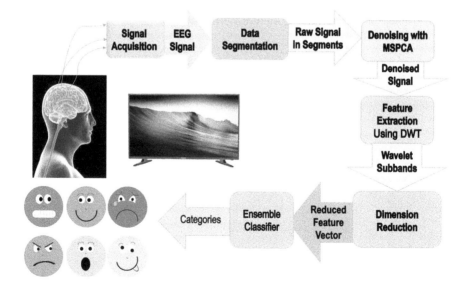

FIGURE 6.1   EEG-based emotion recognition framework.

Pachori, 2014). A significant number of researchers have endeavored, through the extraction of characteristics from EEG data, to categorize emotional states. At the moment, a variety of feature extraction strategies have been suggested as potential solutions for human emotion identification. DWT is extensively utilized method for feature extraction (Murugappan et al., 2010). In addition, ensemble learning algorithms and machine learning algorithms are widely used to classify emotion types from EEG signals (Subasi et al., 2021).

In this study, we combined multiple approaches for classifying emotion types as positive, negative, or neutral. Before classifying, we diminished noise by using MSPCA from EEGs and extracted features via DWT and reduced dimension. After all these stages, we performed the classification process through 10 different classifiers and three different ensemble learning algorithms. The findings that were obtained from Table 6.1 and 6.2 indicate that the SVM with MultiBoosting ensemble approach is the classifier that has the best result, while the naive Bayes classifier is the one that has the poorest result. We calculated F, AUC, and kappa to methodically evaluate the emotion recognition accuracy of the recommended strategy and tested many well-established methodologies to demonstrate that the recommended approach is successful.

This study shows that combining DWT with MultiBoosting and the SVM classifier was highly applicable for EEG-based emotion identification. These findings unequivocally demonstrate that the conceptual framework that was suggested for EEG-based emotion recognition is successful, 3.3% better than with ANN. The following is a list of the advantages of using this strategy:

- All approaches suggested are based on mathematics.
- The proposed method is straightforward and solves signal processing issues.
- The DWT-based feature extraction approach achieves good classification rates. The extracted features are distinct in this case. EEG-based emotion recognition is proposed.

The following is a list of the disadvantages of using this process:

- We only investigated three emotion categories, positive, negative, or neutral.
- We did not apply our method to a larger or more complex data set.
- We used three distinct ensemble learning algorithms but did not compare them with each other.

6.4.2 Performance Evaluation

Numerous performance measures can be used to evaluate the predictive performance of a classification technique. Given that the classification model is trained on a training set, which may be a small subset of the domain, the approximation quality is dependent on the model's generalization properties. For any live performance, it is essential to distinguish between its cost for a specific dataset, especially the training set, and its expected performance on the entire domain (Cichosz, 2014).

The tutoring performance of a model is determined by evaluating it against the training set that was employed to create it. Despite the fact that the objective

TABLE 6.1
Accuracy of Ensemble Classifiers

	Adaboost				MultiBoost				RandomSubSpace			
	Negative	Neutral	Positive	Average	Negative	Neutral	Positive	Average	Negative	Neutral	Positive	Average
SVM	0.981	0.921	0.939	0.947	0.981	0.924	0.946	0.95	0.977	0.914	0.944	0.945
k-NN	0.952	0.765	0.773	0.83	0.934	0.728	0.763	0.808	0.96	0.758	0.791	0.836
ANN	0.985	0.857	0.928	0.923	0.978	0.87	0.904	0.917	0.983	0.891	0.94	0.938
NB	0.983	0.373	0.125	0.494	0.983	0.368	0.154	0.502	0.98	0.348	0.132	0.487
RF	0.967	0.893	0.929	0.93	0.967	0.893	0.929	0.93	0.964	0.874	0.933	0.924
CART	0.952	0.835	0.852	0.88	0.948	0.847	0.872	0.889	0.943	0.84	0.861	0.881
C4.5	0.963	0.86	0.88	0.901	0.965	0.845	0.883	0.898	0.961	0.848	0.879	0.896
REP Tree	0.96	0.842	0.849	0.884	0.951	0.84	0.853	0.881	0.951	0.812	0.869	0.877
Random Tree	0.876	0.71	0.739	0.775	0.876	0.71	0.739	0.775	0.952	0.858	0.847	0.886
LAD Tree	0.953	0.844	0.868	0.888	0.928	0.813	0.872	0.871	0.919	0.776	0.814	0.836

TABLE 6.2

F-Measure, AUC, and Kappa Statistics for Ensemble Classifiers

	F-Measure			AUC			Kappa		
	Adaboost	MultiBoost	RSM	Adaboost	MultiBoost	RSM	Adaboost	MultiBoost	RSM
SVM	0.947	0.95	0.945	0.989	0.989	0.988	0.9205	0.9255	0.9175
KNN	0.829	0.808	0.835	0.892	0.909	0.7545	0.745	0.7125	0.7545
ANN	0.923	0.917	0.938	0.967	0.971	0.99	0.885	0.876	0.907
NB	0.428	0.442	0.422	0.789	0.792	0.826	0.2405	0.2525	0.23
RF	0.93	0.93	0.923	0.986	0.986	0.986	0.8945	0.8945	0.8855
CART	0.88	0.889	0.881	0.971	0.974	0.973	0.8195	0.8335	0.822
C4.5	0.901	0.897	0.896	0.979	0.978	0.975	0.8515	0.8465	0.844
REP Tree	0.884	0.881	0.877	0.972	0.973	0.971	0.8255	0.822	0.816
Random Tree	0.775	0.775	0.885	0.831	0.831	0.972	0.6625	0.6625	0.8285
LAD Tree	0.888	0.871	0.836	0.975	0.964	0.949	0.8325	0.8065	0.7545

of classification techniques is not to classify training data, they are often effective for better comprehending the model and evaluating the performance of the applied classification formula. A model's verifiable performance is its expected performance across the total domain. This demonstrates the model's prognostic utility, i.e., its ability to correctly classify new instances from the specified domain. Since truth category identifiers are typically unavailable for the domain, actual performance is perpetually unknown and may be determined solely by dataset performance. To assess verifiable performance, acceptable analysis procedures are required to reliably estimate the unknown values of the adopted performance measures across the entire domain, which consists primarily of previously unobserved instances. The two most important aspects of classification model analysis are performance measures and analysis procedures (Cichosz, 2014).

K-fold cross-validation involves the random partitioning of dataset X into K sections of equal size, denoted as X_i where i ranges from 1 to K. In order to produce all possible pairs, a single part out of the K available parts is designated as the validation set, while the remaining K-1 parts are combined to create the training set. This operation is performed K times, with each iteration excluding a distinct one of the K components. As the value of K increases, the proportion of training instances also increases, leading to a greater number of robust estimators. However, this results in a decrease in the size of the validation set.

In K-fold cross-validation, the value of K directly impacts the level of increase in performance. As the value of N increases, a smaller K may be used. Conversely, if N is small, a larger K is necessary to facilitate sufficiently large training sets. An example of K-fold cross-validation that is particularly rigorous is the leave-one-out approach. In this method, a dataset consisting of N cases is utilized, with only one case being excluded as the validation set. The remaining N−1 cases are then employed for training purposes. Subsequently, N disconnection pairs are obtained through the omission of an additional instance during each iteration. This is commonly employed in domains such as medical diagnosis, wherein obtaining labeled data is a challenging task. Leave-one-out does not allow for any stratification (Alpaydin, 2014).

A variety of classification measures, particularly for two-class issues, have been planned. There are four possible scenarios. If the prognosis is additionally positive for a positive example, this is frequently a true positive; if the prediction is negative for a positive example, this is frequently a false negative. For a negative example, if the prediction is also negative, we have a true negative; however, if we predict a negative example as positive, we have a false positive. In some two-class problems, we distinguish between the two classes and, consequently, the two types of errors, false positives and false negatives (Alpaydin, 2014).

In several implementations, it can be difficult to determine how frequently the assessed model is incorrect or how expensive its errors are on average; models frequently fail to predict specific categories accurately. This is especially true when categories of the target construct have varying degrees of certainty or incidence rates, which is extremely frequent in sensible classification tasks. This may coincide with the heterogeneous misclassification costs mentioned above. In such circumstances, model performance can be evaluated more thoroughly based on the confusion matrix.

This creates ideal election abundant more durable, though, since there's no singular normal to measure confident models one can be ordered to gather from. Any tests are steered that attempt to provide two corresponding symbols into a singular one, to market this duty. One well-known example is that the F-measure outlined because the harmonious mean of the accuracy and recall:

$$Accuracy = \frac{TP + TN}{TP + FP + FN + TN} \tag{6.3}$$

$$Sensitivity = \frac{TP}{TP + FN} \tag{6.4}$$

$$Specificity = \frac{TN}{FP + TN} \tag{6.5}$$

$$Precision = \frac{TP}{TP + FP} \tag{6.6}$$

$$Recall = \frac{TP}{TP + FN} \tag{6.7}$$

$$F - measure = \frac{2 * precision * recall}{precision * recall} \tag{6.8}$$

The ROC curve is a plot that shows the performance of a twofold variable as its sill is varied. It is constructed by plotting the allergy versus (1 – the specificity) at several threshold settings. A ROC curve, therefore, shows the trade-off between true positive and false positive. A true positive (sensitivity) is a useful outcome, and a false positive (1 – specificity) is an expensive outcome. The ROC curve for a weak model would be an approximately right line connecting these two points.

In contrast, a great model has a well-rounded curve around the space of being left angular. The proportions of the human with every target value affect what the ROC curve will show a good model. ROC analysis can be applied to improve selecting the best models, as the good model will provide points above the diagonal line describing good classification results. The selection of models is normally carried out independently from the cost considerations, and ROC analysis can also be applied to compare models with alike costs, and this thing leads to the sound of decision making (Ahlemeyer-Stubbe & Coleman, 2014).

The kappa statistic takes into account this expected figure by subtracting it from the predictor's successes and expressing the result as a percentage of the total for an optimal predictor. The utmost value of kappa is one hundred percent, and the average value of a random predictor with the same column totals is zero. In conclusion, the kappa statistic is used to measure the agreement between predicted and determined categorizations of a dataset while accounting for chance agreement. However, similar to the simple success rate, it does not account for costs (Hall et al., 2011).

6.4.3 EXPERIMENTAL RESULTS

In this study, we made use of SEED, which is available to the general public and can be accessed from any location. The signals produced by the participants' EEGs were recorded as they watched emotional movie clips while wearing an EEG apparatus designed to assess emotions. The participants wore the device while they watched the emotional movie clips. As described, they also completed a questionnaire to complete on their thoughts and feelings about the films. We classified feelings into one of three primary groups: positive, negative, and neutral using three ensemble learning techniques, which are more sophisticated versions of machine learning algorithms: Adaboost, MultiBoosting, and RSM. However, in order to proceed to the classification stage, dimension reduction, and feature extraction have to be completed first. At this point, the most useful methods for us were DWT and MSPCA, and we benefited from both of them.

We used different machine learning algorithms while using ensemble learning: SVM, K-NN, ANN, naive Bayes, random forest, CART, C4.5, REP tree, random tree, and LAD tree. Using these, we obtained accuracy rates for all three different classes. We also calculated the average accuracy rate for each classifier. When Table 6.1 was examined, we saw that the ensemble learning algorithm with the highest average accuracy rate was MultiBoosting and the SVM classifier. When we examined it on a class basis, we could say that positive and negative emotions could be better distinguished than neutral feelings.

According to Table 6.1, Adaboost gave an average accuracy rate of 94.7% when SVM was the classifier. Here, negative emotion was clearly distinguished with 98.1%, while the accuracy rate was below 94% in other classes. SVM is the best classifier at this stage. In contrast, naïve Bayes showed the worst performance, with an average accuracy of 49.4%. The rates for the other classifiers were as follows: k-NN was 83%, ANN was 92.3%, random forest was 93%, CART was 88%, C4.5 was 90.1%, REP tree was 88.4%, random tree was 77.5%, LAD tree was 88.8%.

MultiBoost with SVM obtained a maximum average accuracy rate of 95% in this study, with 98.1% accuracy with negative emotions; positive and neutral emotions did not obtain the same success. Nevertheless, SVM achieved top performance on these emotions over than 92.3% accuracy rate. The average accuracy rates for the other classifiers were as follows: k-NN was 80.8%, ANN was 91.7%, naïve Bayes was 50.2%, random forest was 93%, CART was 88.9%, C4.5 was 89.9%, REP tree was 88.1%, random tree was 77.5%, and LAD tree was 87.1%.

SVM was also the best classifier with RSM, with a 94.5% accuracy rate, and naïve Bayes was the worst again. The other rates were as follows: k-NN was 83.6%, ANN was 93.8%, naïve Bayes was 48.7%, random forest was 92.4%, CART was 88.1%, C4.5 was 89.6%, REP tree was 87.7%, random tree was 88.6%, and LAD tree was 83.6%. In addition to the accuracy rates, we calculated F1, area under the curve (AUC), and kappa, which show whether the data is balanced and the classifier is consistent. Table 6.2 shows these metrics.

Table 6.2 reflects similar findings for F-measure, AUC, and kappa, and SVM was also the best classifier for all ensemble algorithms. However, SVM obtained the highest F, 0.95, and kappa, 0.9255, via the MultiBoost algorithm. For AUC, AdaBoost and MultiBoost had the same rate. Nonetheless, it didn't change that the MultiBoost algorithm and SVM were the best.

6.5   CONCLUSION

Using EEG data, we classified emotion types based on DWT and MultiBoosting ensemble learning. The suggested approach uses MSPCA for diminishing noises, DWT for extraction of features, descriptive statistics for dimensionality reduction, and 10 commonly used machine learning algorithms for classification. The suggested DWT-based system, employing MultiBoosting+SVM ensemble classifier, obtained a 95% accuracy rate. In conclusion, this research presents a highly accurate approach to analyzing EEG signals in order to recognize emotions. The suggested method is low in mathematical complexity, and automating the process removes the need for any complex techniques that can improve the performance of classification.

For future works, biological signals can be analyzed with this approach to recognize other classes. We intend to test the performance of the proposed strategy on larger and more varied datasets. Additionally, we only tested the identification of three emotions in the present study. It would be extremely informative to test the suggested DWT-based MultiBoosting+SVM approach using a broader range of emotions.

6.6   FUNDING

This work was supported by Effat University with the Decision Number of UC#7/28 Feb. 2018/10.2–44i, Jeddah, Saudi Arabia.

REFERENCES

Ahlemeyer-Stubbe, A., & Coleman, S. (2014). *A practical guide to data mining for business and industry*. John Wiley & Sons.

Alpaydin, E. (2014). *Introduction to machine learning*. MIT Press.

Alsberg, B. K., Woodward, A. M., Winson, M. K., Rowland, J. J., & Kell, D. B. (1998). Variable selection in wavelet regression models. *Analytica Chimica Acta, 368*(1), 29–44.

Bajaj, V., & Pachori, R. B. (2014). Human emotion classification from EEG signals using multiwavelet transform. *2014 International Conference on Medical Biometrics*. IEEE Shenzen China, 125–130.

Bakshi, B. R. (1998). Multiscale PCA with application to multivariate statistical process monitoring. *AIChE Journal, 44*(7), 1596–1610.

Bakshi, B. R. (1999). Multiscale analysis and modeling using wavelets. *Journal of Chemometrics, 13*(3–4), 415–434.

Bakshi, B. R., Bansal, P., & Nounou, M. N. (1997). Multiscale rectification of random errors without fundamental process models. *Computers & Chemical Engineering, 21*, S1167–S1172.

Bauer, E., & Kohavi, R. (1999). An empirical comparison of voting classification algorithms: Bagging, boosting, and variants. *Machine Learning, 36*(1–2), 105–139.

Bhatti, A. M., Majid, M., Anwar, S. M., & Khan, B. (2016). Human emotion recognition and analysis in response to audio music using brain signals. *Computers in Human Behavior, 65*, 267–275.

Chai, X., Wang, Q., Zhao, Y., Li, Y., Liu, D., Liu, X., & Bai, O. (2017). A fast, efficient domain adaptation technique for cross-domain electroencephalography (EEG)-based emotion recognition. *Sensors, 17*(5), 1014.

Chakladar, D. D., & Chakraborty, S. (2018). EEG based emotion classification using "correlation based subset selection." *Biologically Inspired Cognitive Architectures, 24*, 98–106.

Cichosz, P. (2014). *Data mining algorithms: Explained using R*. John Wiley & Sons.

Daimi, S. N., & Saha, G. (2014). Classification of emotions induced by music videos and correlation with participants' rating. *Expert Systems with Applications, 41*(13), 6057–6065.

Freund, Y., & Schapire, R. E. (1995). A desicion-theoretic generalization of on-line learning and an application to boosting. *European Conference on Computational Learning Theory*. Berlin, Heidelberg: Springer Berlin Heidelberg, 23–37.

Ghaemi, A., Rashedi, E., Pourrahimi, A. M., Kamandar, M., & Rahdari, F. (2017). Automatic channel selection in EEG signals for classification of left or right hand movement in brain computer interfaces using improved binary gravitation search algorithm. *Biomedical Signal Processing and Control, 33*, 109–118.

Ghorbanian, P., Devilbiss, D., Simon, A., Bernstein, A., Hess, T., & Ashrafiuon, H. (2012). Discrete wavelet transform EEG features of Alzheimer'S disease in activated states. *Annual International Conference of the IEEE Engineering in Medicine and Biology Society*. San Diego, California, USA, 2937–2940.

Grąbczewski, K. (2014). *Meta-learning in decision tree induction* (Vol. 1). Springer.

Hall, M., Witten, I., & Frank, E. (2011). *Data mining: Practical machine learning tools and techniques*. Kaufmann.

Hancer, E., & Subasi, A. (2022). EEG-based emotion recognition using dual tree complex wavelet transform and random subspace ensemble classifier. *Computer Methods in Biomechanics and Biomedical Engineering*, 1–13.

Hemanth, D. J., & Anitha, J. (2018). Brain signal based human emotion analysis by circular back propagation and Deep Kohonen Neural Networks. *Computers & Electrical Engineering, 68*, 170–180.

Ho, T. K. (1998). The random subspace method for constructing decision forests. *IEEE Transactions on Pattern Analysis and Machine Intelligence, 20*(8), 832–844.

Iacoviello, D., Petracca, A., Spezialetti, M., & Placidi, G. (2015). A real-time classification algorithm for EEG-based BCI driven by self-induced emotions. *Computer Methods and Programs in Biomedicine, 122*(3), 293–303.

Katsigiannis, S., & Ramzan, N. (2018). Dreamer: A database for emotion recognition through EEG and ECG signals from wireless low-cost off-the-shelf devices. *IEEE Journal of Biomedical and Health Informatics, 22*(1), 98–107.

Kumar, N., Khaund, K., & Hazarika, S. M. (2016). Bispectral analysis of EEG for emotion recognition. *Procedia Computer Science, 84*, 31–35.

Lahane, P., & Sangaiah, A. K. (2015). An approach to EEG based emotion recognition and classification using kernel density estimation. *Procedia Computer Science, 48*, 574–581.

Liu, Y.-J., Yu, M., Zhao, G., Song, J., Ge, Y., & Shi, Y. (2018). Real-time movie-induced discrete emotion recognition from EEG signals. *IEEE Transactions on Affective Computing, 9*(4), 550–562.

Murugappan, M., Ramachandran, N., & Sazali, Y. (2010). Classification of human emotion from EEG using discrete wavelet transform. *Journal of Biomedical Science and Engineering, 3*(4), 390.

Othman, M., Wahab, A., Karim, I., Dzulkifli, M. A., & Alshaikli, I. F. T. (2013). EEG emotion recognition based on the dimensional models of emotions. *Procedia-Social and Behavioral Sciences, 97*, 30–37.

Petrantonakis, P. C., & Hadjileontiadis, L. J. (2011). A novel emotion elicitation index using frontal brain asymmetry for enhanced EEG-based emotion recognition. *IEEE Transactions on Information Technology in Biomedicine, 15*(5), Article 5.

Pham, T. D., Tran, D., Ma, W., & Tran, N. T. (November 2015). Enhancing performance of EEG-based emotion recognition systems using feature smoothing. *International Conference on Neural Information Processing.* Istanbul, Turkey, 95–102.

Quinlan, J. R. (July 1996). Bagging, boosting, and C4. 5. *Proceedings of the AAAI Conference on Artificial Intelligence.* Washington, DC, 725–730.

Skurichina, M., & Duin, R. P. (2002). Bagging, boosting and the random subspace method for linear classifiers. *Pattern Analysis & Applications, 5*(2), 121–135.

Su, Y., Chen, P., Liu, X., Li, W., & Lv, Z. (2018). A spatial filtering approach to environmental emotion perception based on electroencephalography. *Medical Engineering & Physics, 60*, 77–85.

Subasi, A., Tuncer, T., Dogan, S., Tanko, D., & Sakoglu, U. (2021). EEG-based emotion recognition using tunable Q wavelet transform and rotation forest ensemble classifier. *Biomedical Signal Processing and Control, 68*, 102648.

Subramanian, R., Wache, J., Abadi, M. K., Vieriu, R. L., Winkler, S., & Sebe, N. (2018). ASCERTAIN: Emotion and personality recognition using commercial sensors. *IEEE Transactions on Affective Computing, 9*(2), 147–160.

Trygg, J., & Wold, S. (1998). PLS regression on wavelet compressed NIR spectra. *Chemometrics and Intelligent Laboratory Systems, 42*(1–2), 209–220. https://doi.org/10.1016/S0169-7439(98)00013-6

Tuncer, T., Dogan, S., & Subasi, A. (2021). LEDPatNet19: Automated emotion recognition model based on nonlinear LED pattern feature extraction function using EEG signals. *Cognitive Neurodynamics.* https://doi.org/10.1007/s11571-021-09748-0

Verma, G. K., & Tiwary, U. S. (2014). Multimodal fusion framework: A multiresolution approach for emotion classification and recognition from physiological signals. *NeuroImage, 102*, 162–172.

Wang, X.-W., Nie, D., & Lu, B.-L. (2014). Emotional state classification from EEG data using machine learning approach. *Neurocomputing, 129*, 94–106.

Webb, G. I. (2000). Multiboosting: A technique for combining boosting and wagging. *Machine Learning, 40*(2), 159–196.

Xu, T., Zhou, Y., Wang, Z., & Peng, Y. (2018). Learning emotions EEG-based recognition and brain activity: A survey study on BCI for intelligent tutoring system. *Procedia Computer Science*, *130*, 376–382.

Yang, Y., Wu, Q. J., Zheng, W.-L., & Lu, B.-L. (2018). EEG-based emotion recognition using hierarchical network with subnetwork nodes. *IEEE Transactions on Cognitive and Developmental Systems*, *10*(2), 408–419.

Yin, Z., Zhao, M., Wang, Y., Yang, J., & Zhang, J. (2017). Recognition of emotions using multimodal physiological signals and an ensemble deep learning model. *Computer Methods and Programs in Biomedicine*, *140*, 93–110.

Zhang, Y., Ji, X., & Zhang, S. (2016). An approach to EEG-based emotion recognition using combined feature extraction method. *Neuroscience Letters*, *633*, 152–157.

Zhao, G., Ge, Y., Shen, B., Wei, X., & Wang, H. (2018). Emotion analysis for personality inference from EEG signals. *IEEE Transactions on Affective Computing*, *9*(3), 362–371.

Zheng, W.-L., & Lu, B.-L. (2015). Investigating critical frequency bands and channels for EEG-based emotion recognition with deep neural networks. *IEEE Transactions on Autonomous Mental Development*, *7*(3), 162–175.

Zheng, W.-L., Zhu, J.-Y., & Lu, B.-L. (2017). Identifying stable patterns over time for emotion recognition from EEG. *IEEE Transactions on Affective Computing*, *10*(3), 417–429.

Zheng, W.-L., Zhu, J.-Y., & Peng, Y. (September 2014). Bao-Liang Lu. EEG-based emotion classification using deep belief networks. *2014 IEEE International Conference on Multimedia and Expo (ICME)*. Chengdu, China, 1–6.

7 Automatic Emotion Detection by EEG Analysis Using Graph Signal Processing

Ramnivas Sharma and Hemant Kumar Meena

7.1 INTRODUCTION

The evaluation of cognitive functions and clinical state of patients is a crucial aspect of delivering e-health care and developing novel human-machine interfaces. Emotions play a significant role in affecting cognition, and therefore, detecting emotions through analyzing physiological signals is essential. Among various physiological signals, electroencephalography (EEG) signals have become a preferred choice due to their simplicity and acceptability among subjects.

In this chapter, we evaluate the findings of current studies and talk about how to conceptualize fundamental emotions as well as feature extraction, selection, and classification strategies, concluding with a discussion of the challenges and opportunities facing researchers exploring EEG-based emotion analysis systems. These signals provide rich and meaningful data with high temporal resolution that can be accessed through inexpensive, portable EEG devices, but despite the promising results yielded by EEG-based emotion recognition systems, there are many challenges that must be overcome. Therefore, it is essential to determine how brain signals can be used to estimate emotions, taking into account various factors such as sampling frequency, number of subjects, electrode location, and nature of emotion-eliciting stimulation [1, 2]. Beginning with an overview to the concepts of emotion classification, different kinds of EEG electrodes, and the relationship between brain waves and emotion analysis, we examine and compares previous works on emotion classification using EEG data. The following sections of the chapter discuss stimulation materials, emotion-related databases, EEG recording equipment, EEG preprocessing methods, and modern analysis approaches such as feature extraction, selection, and classification [3, 4].

The identification of neural patterns related to various emotional states is a developing area of study known as emotion recognition using EEG signals. Using electrodes positioned on the scalp, the noninvasive neuroimaging technique EEG measures electrical activity generated by the brain. Researchers can analyze these signals to identify different emotional states, such as happiness, sadness, anger, or fear, by detecting patterns of brain activity. Emotion recognition using EEG signals

DOI: 10.1201/9781003479970-7

has potential applications in several areas, including e-health care, human–computer interaction (HCI), and psychology. EEG-based emotion recognition could be used to enhance clinical diagnoses and monitor treatment outcomes in e-health care. In terms of HCIs, more intelligent and responsive interfaces such as virtual assistants or smart homes that can adjust to users' emotional states could be developed with EEG-based emotion recognition. In psychology, EEG can help to better understand the neural mechanisms underlying emotional states and to develop more effective therapies for mental health disorders [5, 6].

Yet employing EEG for emotion identification comes with a number of difficulties. One of the main challenges is the variability of emotional states across individuals. People experience and express emotions differently, which can make it difficult to identify consistent patterns of brain activity that correspond to specific emotions. Additionally, various factors can affect EEG signals such as the time of day, underlying mental conditions, language, and cultural differences. Despite these challenges, recent studies have shown promising results in using EEG signals to detect emotions.

Researchers have developed various techniques and methodologies for analyzing EEG signals such as feature extraction, selection, and classification that can help to identify consistent patterns of brain activity associated with different emotional states [5, 6]. Furthermore, advances in EEG technology such as portable and wireless EEG devices have made it easier to collect EEG data in naturalistic settings and to study the neural correlates of emotions in real-life situations. Hence, emotion recognition using EEG signals is an exciting and rapidly growing field of research with potential applications in various fields (see Figure 7.1). Despite the challenges associated with using EEG for emotion recognition, recent studies have shown promising results, and researchers continue to develop new methods and techniques to improve the accuracy and reliability of emotion recognition using EEG.

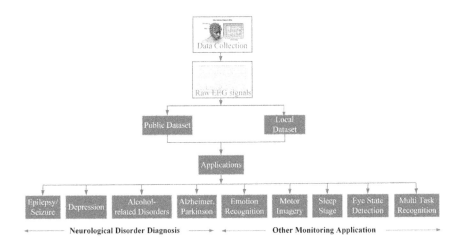

FIGURE 7.1 Different applications of EEG signals.

Recognizing emotions can go beyond identifying facial expressions [1–3] or vocal cues, as physiological signals like EEG, electrocardiogram (ECG), electromyogram (EMG), skin temperature (ST), respiration rate (RT), and galvanic skin response (GSR) [4, 5] can also indicate emotional states. However, while ST and GSR can measure only arousal levels and EMG can measure valence, EEG signals can detect actual emotions, particularly through gamma and beta waves. For instance, gamma wave signals have been linked to happiness and sadness [6], while beta and gamma bands [7] can describe valence and arousal.

Nonetheless, different regions of the brain can also be involved in expressing emotions, and the nonlinear and nonstationary nature of EEG signals requires feature extraction and analysis. Examining these characteristics in the time, frequency, time–frequency, and spatial domains is possible. Many methods, such as statistical and Hjorth parameters [8–11], nonstationary index, higher-order crossover, and power spectrum analysis and nonlinear properties such fractal dimension and entropy, have been employed to analyze emotions with EEG signals, yielding significant results in recent studies [12–14]. Emotion is a crucial aspect of human physiology that affects physical, mental, and social well-being. Humans experience emotions in various ways, consciously or subconsciously [15], in response to external stimuli. However, emotions are often misunderstood as mood or affect. Depression, anxiety, and bipolar disorders are some of the emotional problems that people face due to a busy modern lifestyle and other factors.

Affective computing is an innovative field that seeks to enhance the naturalness of human–computer interaction. Two models can be used to describe emotions: the discrete emotions model and the two-dimensional valence–arousal model [16]. Researchers usually rely on physiological signals or emotional expressions from users to recognize emotional states, which can include blood volume pulse, skin temperature, EMG, ECG, GSR, and EEG [17]. EEG signals can be divided into five frequency bands, including delta (δ), theta (θ), alpha (α), beta (β), and gamma (γ), that correspond to different mental states. Emotion recognition can be performed using either a discrete or a dimensional model. The discrete model proposed by Ekman identifies six basic emotions: fear, sadness, disgust, anger, happiness, and surprise. In contrast, Russell developed a two-dimensional bipolar model of valence and arousal [18].

7.2   RELATED WORK

The EEG emotion categorization problem and various disease detection difficulties have been addressed using a number of different methods in the literature. Many well-known methods, including CNN [19, 20] and deep neural network [20] algorithms, have been used to address a variety of characterization issues, picture classification [21, 22], including segmentation, object detection [23], and tracking [24]. Machine learning (ML) techniques have been used to address a wide range of complex problems in a variety of industries, including the medical, financial, and other professions.

In this study, we investigated the use of ML in medicine, with an emphasis on common technologies and how they affect medical diagnosis. In important fields like cancer, pharmaceutical applications, the brain, emotion classification, medical imaging, and monitoring devices, ML models are routinely used [25–27]. Alhalaseh et al. [28] focused on a system that effectively combines the required steps of EEG signal processing and feature extraction and used the Higuchi fractal dimension and entropy for the feature extraction. We used the DEAP database to classify emotions using naive Bayes, k-nearest neighbor (KNN), CNN, and decision trees (DT). Among others, Moshfeghi [29] monitored the frontal and central lobes of the brain using six electrodes to collect EEG data. They used Waikato Environment for Knowledge Analysis machine learning was used to divide the collected variables into three groups with tenfold replication. The accuracy result for categorizing using three emotional states was 54%, whereas the accuracy result for categorizing using just two emotional states was 74%.

Joshi et al. [30] used bilinear long short-term memory network decoding and linear conceptualization of differential entropy feature selection. The suggested method used the SEED database to separate emotions into positive, neutral, or negative. It also used DEAP, or the database for emotion analysis using physiological signals, to classify emotions according to their valence and arousal. The classification of emotions has improved in terms of prediction accuracy. Demir et al. [31] used wavelet transform and continuous wavelet transform to convert the EEG signals into EEG rhythm images during the preprocessing stage. These photos were afterward fed into different CNN models as input. The findings demonstrate that Alex Net functionalities with an alpha rhythm generate higher valence discrimination accuracy ratings than other deep feature functionalities with MobilNetv2 functionalities having the greatest arousal discrimination accuracy score (98.93% in delta rhythm).

EEG signals from multiple subjects were compared using a profitable machine learning-based methodology developed by Placidi et al. [32] for the DEAP dataset. Principle component analysis (PCA) was used to select a small subset of features from the recommended five sub-trials based on their averaged values. These features were then applied to the SVM. Rahman et al. [33] combined PCA with the t statistic for feature extraction. Their work also aids in the process of lowering signal dimensions and choosing more dependable characteristics for t-statistic-based feature extraction. The aforementioned proposed technique was tested on the SEED dataset with four classifiers, SVM, LDA, ANN, and KNN, and the results showed that ANN and SVM performed better, with accuracy rates of 84.3% and 77.1%, respectively.

7.3   DIFFERENT HUMAN EMOTIONS AND EEG

Emotions have an impact on a person's psychological and physical health. There are many different emotions, such as anxiety, anger, fear, hatred, joy, sorrow, pleasure, and boredom. Facial expressions, signing, and talking are all ways that people can communicate their emotions, all of which improve interpersonal relationships [34, 35]. There are two schools of thought on emotions: One regards them as physiological interactions, while the other sees them as natural states of people [36, 37]. Two approaches, the dimension method and discrete fundamental emotion classification, can be used to categorize emotions [19, 38].

Emotions are essentially positive or negative in valence. Using the dimension approach, emotions are divided into three groups according to their valence, arousal, and dominance [20, 39]. Valence describes how strongly a feeling is favorable or negative. As a result, it is further classified into two categories: high valence or positive valence emotions are glad, calm, delighted, enthusiastic, etc., whereas low valence or negative valence emotions are fear, tenseness, fury, sadness, boredom, and so on [39, 40].

The term "arousal" [39] refers to the intensity of an emotional state and reflects the degree of zeal, eagerness, or apathy. Additionally, it is split into two groups: Negative low-arousal feelings include drowsiness, boredom, calmness, and melancholy, and negative high-arousal emotions include anger, excitement, anxiety, vigor, and joy. Figure 7.2 shows how Russell's 16 emotions, which span from calm to stimulated, can be arranged on a two-dimensional plane based on arousal (from pleasant to unpleasant). The cerebrum, which is divided into the left and right hemispheres, is the most essential and vital component of the human brain. Each hemisphere is made up of the frontal, temporal, parietal, and occipital lobes. The frontal lobe is in charge of controlling things like personality, attention, and physical movement. In addition to sensations like hunger and thirst, the temporal lobe also regulates facial recognition and hearing [40, 41]. Taste, touch, and other senses are managed by the parietal lobe. The occipital lobe helps humans distinguish colors, form, size, depth, and distance [40].

By merging essential emotions, Robert Plutchik created the wheel approach of emotion definitions in 1980 [43]. He also suggested eight other emotions: joy versus melancholy, rage versus fear, trust versus disgust, and surprise versus anticipation. Cowen et al. also identified 27 different emotions using statistical techniques, with 2185 brief movies serving as the emotional stimuli for 800 individuals [44]. Table 7.1 provides a summary of Russell's valence arousal model, which was published in 1980 [45]. In the table, HAPV indicates high arousal positive valence, HANV means high arousal negative valence, LANV indicates low arousal negative valence, and LAPV represents low arousal positive valence.

A Categorical theory

B Dimensional theory

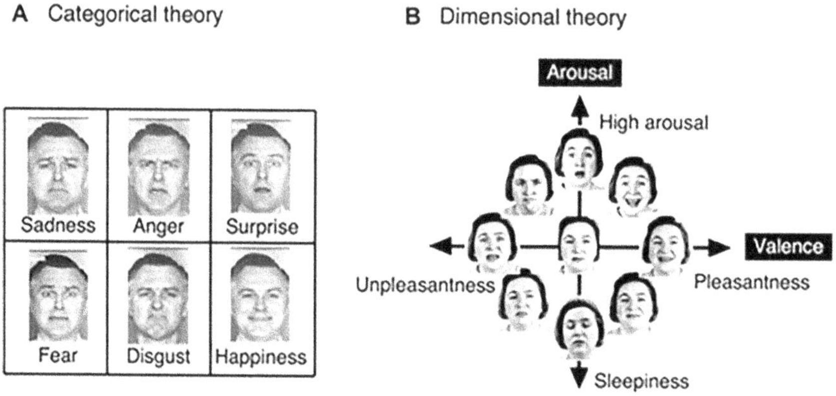

FIGURE 7.2   Emotions model [42].

TABLE 7.1

Emotions on the Valence–Arousal Plane

Position	Emotional state	Emotions
1st quadrant	HAPV	Happy, excited, pleased
2nd quadrant	HANV	Angry, nervous, annoying
3rd quadrant	LANV	Sad, bored, sleepy
4th quadrant	LAPV	Relax, peaceful, calm

7.4 DATASETS

The study of EEG waves can be used to categorize various emotions using a number of accessible datasets. Widely used public datasets for this purpose include DEAP [46], DREAMER [47], IAPS [48], ASCERTAIN [49], MANHOB-HCI [50], SEED [51], AMIGOS [52], CAPS [53], and INTERFACES [54]. Table 7.2 provides a list of the datasets utilized to examine emotional states. These datasets were used in the studies that we looked at to describe and identify common emotions including joy, rage, fear, surprise, sadness, and disgust. Using various rating scales, three datasets calculated the valence and arousal of each (1–9 for DEAP and AMIGOS, 1–5 for DREAMER). Positive, negative, and neutral emotions were examined in two datasets, MELD and SEED. Few researchers have focused on positive versus negative emotions.

In the investigations, stimuli that are significant to the analysis elicit emotional states. Some stimuli generate variations in EEG signals when emotional states vary, making them useful for assessing emotional status. The most frequent methods for evoking emotions include images, audiovisual materials, music, recollections, self-induction, and games. IAPS and CAPS dataset images have been employed as stimuli in numerous investigations [55, 56]. The stimuli's durations ranged from 35 to 393 seconds. We examined 54 studies for this review, and 33 of them employed audio-video clips as stimuli while 9 used videos. The remaining 12 studies examined how to induce emotional reactions using visuals and music that lasted between 15 and 60 seconds. Because audio-video clips contain both scenes and music, as well as expose participants to more realistic circumstances, some researchers have found that they are far more effective at evoking emotions than other stimuli.

7.5 DIFFERENT TECHNIQUES FOR ANALYZING THE EEG SIGNALS

The analysis of EEG signals can be carried out in four phases, i.e., preprocessing, feature extraction, postprocessing, and result analysis [57]. These steps are shown in Figure 7.3. In a deep neural network, feature extraction and feature selection are not required; raw signals are directly passed for classification and analysis purpose to the result analysis phase. It means we can ignore the first three stages of the preprocessing mechanism. In addition, one can avoid these steps if we used already preprocessed data, mainly some publicly available datasets such as DEAP and SEED.

TABLE 7.2

Descriptions of Public Datasets

Name	Participants	Stimulus	Emotions
SEED	15	15 video clips	positive, neutral, negative
IAPS	1483	pictures	arousal, valence, dominance
CAPS	46	852 pictures	arousal, valence, dominance
DREAMER	23	18 video clips	valence, arousal, dominance
AMIGOS	40	20 video clips	arousal, valence, dominance, liking, familiarity; discrete/basic emotions
ASCERTAIN	58	36 video clips	arousal, valence
DEAP	32	40 video clips	valence, arousal, liking, dominance, familiarity

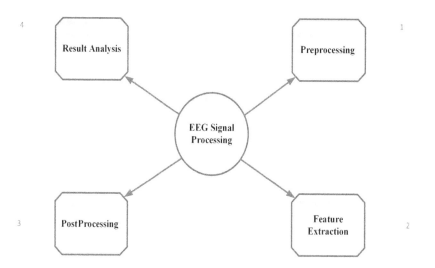

FIGURE 7.3   General approach to analyze EEG signals.

7.5.1 PREPROCESSING

The preprocessing is divided into three steps, i.e., down sampling, artifact removal and feature scaling.

   Downsampling: The recorded data by the metal electrode over the scalp with different EEG devices is down-sampled to different frequencies such as from 512 Hz to 64 Hz, 256 Hz to 16 Hz depending upon the requirement.

   Artifact removal: Artifacts are generated due to errors in experimental settings, environmental noise, physiological signals, or biological artifacts. These include eye movements, muscle activity (or EMG artifacts), and

ECG artifacts. We should remove these artifact types to get optimized results [57].

Feature scaling: Scaling is crucial for the dataset's features to display symmetrical behavior. One of the most commonly used feature scaling techniques is normalization. Some machine learning algorithms' objective functions will not function effectively without normalization since the range of raw data's values fluctuates wildly.

7.5.2 FEATURE EXTRACTION

Feature extraction involves extracting features from the primary signal to obtain consistent categorization. Several methods [58] are used to extract and analyze time, frequency, and time–frequency features. Feature extraction, dimension reduction, or feature selection are used to minimize the initial collection of features if each feature affects categorization [58]. Meaningless or redundant features are removed during feature selection/reduction [59], resulting in faster performance and classification compared with the initial features. An accurate feature extraction approach is required to extract a distinct and valuable set of features from source signals [58]. In other words, if the obtained features do not accurately define the signals used and irrelevant, accuracy of model may suffer.

The classification of neurological disorders and other monitoring applications based on EEG signals is done with the help of feature extraction, which describes in Table 7.3. As EEG, signals contain many data points, distinct and meaningful features are recovered using various feature extraction methods. The most common methods include wavelet transform, power spectral density (PSD), statistical, short-time Fourier transform, wavelet entropy, differential entropy (DE), and empirical mode decomposition. Frequencies and amplitudes can represent signal patterns in EEG signal analysis. The periodogram and the correlogram are two well-known nonparametric techniques. Feature extraction reduces the initial data into a lower dimension, containing the initial vector's practical information. As a result, based on the nature of the collection, it is critical to identify the essential elements that characterize the entire dataset.

7.5.3 POSTPROCESSING

Feature selection and dimensionality reduction are the types of postprocessing. To build a model, one must choose select a subset of pertinent features. Feature selection removes irrelevant variable or noise, reduces computational complexity and overfitting, and increases the generalizability of the model. Dimensionality reduction [60] (sometimes also known as feature extraction) combines features to extract a new smaller set of features. The majority of feature extraction methods produce redundant features [61]. To improve classification accuracy, the characteristics or dimensions must be decreased.

Using feature extraction, data in a higher-dimensional domain is turned into a lower dimensional domain. Data transformations can be linear, like in PCA, but nonlinear dimension reduction approaches are also available [62]. To interpret the

TABLE 7.3
Previous Studies of Emotion Recognition from EEG Signals

Ref.	Datasets	Feature extraction method	Classification method	Emotions	Accuracy
[70]	DEAP and SEED	PSD, DASM, RASM, ASM, DCAU	KNN, LR, SVM, GELM (5-fold cross validation)	positive, negative, neutral	DEAP: 69.67%, SEED: 91.07% (GELM)
[71]	DEAP	DWT (entropy and energy)	KNN	valence and arousal	Valence: 89.54%, Arousal: 89.81%
[72]	Audio-video clips	Sample entropy, Tsallis entropy, FD, Hurst	MC-LS-SVM	happy, fear, sad, relax	Happy: 92.79%, Fear: 87.62%, Sad: 88.98%, Relax: 93.13%
[73]	SEED	DE	DNNs	positive, negative, neutral	Overall Accuracy: 93.28%
[28]	DEAP	Spectral entropy, Higuchi's fractal dimension	CNN, KNN, NB, DT	valence and arousal	Overall accuracy: 95.20% (using CNN)
[74]	DEAP and DREAMR	raw signals, DE, statistical features	RACNN	valence and arousal	96.65: (Valence), 97.11%: (Arousal)
[75]	DEAP and MANHOB-HCI	statistical features, Hjorth parameters, PSD	HFCNN	valence and arousal	Overall; DEAP: 84.71%, MANHOB-HCI: 89%
[76]	Audio-video clips	Hjorth complexity, log energy, DASDV	DT, KNN, SVM	fear, sad, happy, relax	DT: 87.7%, KNN: 93.8%, SVM: 93.1%
[77]	SEED	Spectral entropy, Shannon entropy, temporal entropy	Autoencoder-based random forest	negative, positive, neutral	ARF: 94.4%

data and obtain more precise results, the dimensions of EEG signals need to be reduced. To minimize the number of dimensions, LDA, PCA, and ICA can be used. In order to clarify our grasp of dimension reduction, these three strategies have been defined [62].

7.5.4 RESULT ANALYSIS

Analyzing EEG signals involves multiple steps. The first step is signal preprocessing, where the raw data is cleaned and preprocessed to remove noise, artifacts, and baseline drift. This is followed by feature extraction, where relevant information extracted from the preprocessed EEG signals. This information can be in the form of time, frequency, or time–frequency analysis. The next step is feature classification of the EEG signals (Figure 7.4). This step can be performed using supervised or unsupervised learning techniques. Supervised learning involves training a model with labeled data, while unsupervised learning identifies patterns in the data without prior knowledge of the classes.

Finally, the results of the classification are interpreted to draw conclusions about the underlying brain activity. This can involve identifying the location and timing of specific events or the occurrence of a cognitive process. It is important to acknowledge that the specific techniques used for each step of EEG signal analysis can vary depending on the research question, the characteristics of the EEG signals being analyzed, and the available computational resources. The researcher can use various machine learning algorithms [62] such as supervised, unsupervised, and deep learning neural networks [63] and graph signal processing [64] for EEG signal analysis. With the help of these techniques and EEG signals, one can diagnose neurological disorders such as epilepsy or seizures and monitor other applications such as emotion detection and sleep stage categorization. Supervised learning is used for classification (discrete) and regression (continuous) and unsupervised learning for clustering and dimensionality reduction.

In addition to machine learning, deep learning algorithms are utilized to predict depressive subjects and brain disorders with EEG signals. Long short-term memory (LSTM) is primarily applied to prediction tasks among all deep learning techniques. For additional chances to obtain the highest accuracy, the author of [39, 65] used two deep learning structures and applied them to the same prepared dataset. The advantages of the LSTM architecture's memorizing capabilities [65] and the built-in feature extraction capability of CNN constructions allowed combined models with convolution neural network structure and LSTM blocks to achieve better results than other models.

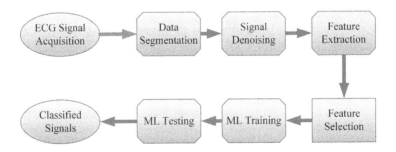

FIGURE 7.4 Basic structure to analyze EEG signals.

7.6   RESEARCH GAPS AND FUTURE DIRECTIONS

One of the essential questions is how to represent efficiently, handle, analyze, and visualize large-scale structured data, particularly data from complicated domains like networks, Alzheimer's disease fMRI data [66] and graphs, which is one of the significant challenges in existing machine learning techniques. Graph signal processing (GSP) [64], a developing branch of signal processing models and methods aimed at estimating data based on graphs, offers a new route for study to address this issue. In this chapter, we present how GSP concepts and tools, including graph filters and transforms, have aided the development of cutting-edge machine learning approaches. More importantly, we offer a new perspective on the future expansion of GSP development that could act as a link between applied mathematics and signal processing on the one hand and machine learning and social network analysis on the other. The features from the EEG signal obtained in the previous stage are then extracted and turned into a feature vector. Two common issues in the analysis of EEG signals are that most applications have lacked access to larger datasets, limiting the validation of the models for use in real-world applications such as clinical practice and that because real-time data is rife with artifacts, highly efficient filtering is required to help decrease noise and improve model performance. GSP has been established addressing these problems.

7.6.1   GSP FOR EMOTION RECOGNITION USING EEG SIGNALS

Graph signal processing (GSP) [64] is a technique that can be applied to emotion recognition by analyzing the connectivity patterns of brain regions. GSP represents the brain regions as nodes and the connectivity between them as edges in a graph. The emotion recognition task can then be treated as a graph signal classification problem, where the goal is to classify the graph signals (connectivity patterns) into different emotional states.

GSP can also be used to identify the most important brain regions for emotion recognition by analyzing the graph topology. This can help in understanding the neural mechanisms underlying emotions and can lead to the development of more effective treatments for mood disorders. It is important to acknowledge that the use of GSP in emotion recognition is an active area of research and that there are ongoing studies exploring its effectiveness in this domain.

GSP [64, 66] is now an obvious choice for many applications involving brain image processing and disorders. GSP provides tools for analyzing irregular domain data and graph networks. In neuroscience, brain activities are associated with complex functional and structural connectivity networks. GSP is used for handling such data with irregular structures and dynamic and non-Euclidean domains. Examples of such applications are social networks, sensor feeds, online traffic, supply chains, and biological systems. CNNs do not perform well with graph structures [67, 68] or data with irregular domain, the CNN does not perform well, and it will be beyond the scope of CNN [67] analysis. Geometric deep learning is a field of research that extends deep learning to non-Euclidean data. To take into account the merits of GSP, graph convolutional neural network classifiers have been developed in [67–69].

7.7 CONCLUSION

In order to properly examine the many articles on the topic of identifying emotions from EEG data, we thoroughly analyzed them. We gathered pertinent research materials from reputable databases in order to ensure an organized approach. The main goal of this study was to identify the critical procedures required for enhancing EEG signal analysis for emotion estimation. This required the use of common datasets, materials for evoking emotions, EEG equipment, and the evaluation of the effect of artifacts on brain waves.

We suggested a methodical approach for choosing the right number of individuals, stimulation tools, feature extraction strategies, and selection methods. The techniques for precise feature extraction, data dimensionality reduction, and data smoothing were also investigated. To extract valuable features, we suggested a variety of artifact removal methods, including hybrid ones. Future research should concentrate on expanding the participant pool and creating subject-independent characteristics in order to increase accuracy. We concluded that audio-video clips offer the finest representation of emotional situations that occur in real life. Finally, we proposed that deep learning methods, including CNN, DBNs, and RNN, are more effective for evaluating vast amounts of data, and their use is expected to improve accuracy in future research and development.

REFERENCES

1. C. Busso, Z. Deng, S. Yildirim, M. Bulut, C. M. Lee, A. Kazemzadeh, S. Lee, U. Neumann, and S. Narayanan, "Analysis of emotion recognition using facial expressions, speech and multimodal information," in *Proceedings of the 6th International Conference on Multimodal Interfaces*, pp. 205–211, 2004.
2. Y.-T. Matsuda, T. Fujimura, K. Katahira, M. Okada, K. Ueno, K. Cheng, and K. Okanoya, "The implicit processing of categorical and dimensional strategies: An fMRI study of facial emotion perception," *Frontiers in Human Neuroscience*, vol. 7, p. 551, 2013.
3. Y. Miyakoshi and S. Kato, "Facial emotion detection considering partial occlusion of face using bayesian network," in *2011 IEEE Symposium on Computers & Informatics*, pp. 96–101, IEEE, 2011.
4. B. Schuller, G. Rigoll, and M. Lang, "Speech emotion recognition combining acoustic features and linguistic information in a hybrid support vector machine-belief network architecture," in *2004 IEEE International Conference on Acoustics, Speech, and Signal Processing*, vol. 1, pp. I–577, IEEE, 2004.
5. R. Nawaz, K. H. Cheah, H. Nisar, and V. V. Yap, "Comparison of different feature extraction methods for EEG-based emotion recognition," *Biocybernetics and Biomedical Engineering*, vol. 40, no. 3, pp. 910–926, 2020.
6. Y. Zhang, J. Chen, J. H. Tan, Y. Chen, Y. Chen, D. Li, L. Yang, J. Su, X. Huang, and W. Che, "An investigation of deep learning models for EEG-based emotion recognition," *Frontiers in Neuroscience*, vol. 14, p. 622759, 2020.
7. D. Huang, C. Guan, K. K. Ang, H. Zhang, and Y. Pan, "Asymmetric spatial pattern for EEG-based emotion detection," in *The 2012 International Joint Conference on Neural Networks (IJCNN)*, pp. 1–7, IEEE, 2012.
8. T. F. Bastos-Filho, A. Ferreira, A. C. Atencio, S. Arjunan, and D. Kumar, "Evaluation of feature extraction techniques in emotional state recognition," in *2012 4th International Conference on Intelligent Human Computer Interaction (IHCI)*, pp. 1–6, IEEE, 2012.

9. A. T. Sohaib, S. Qureshi, J. Hagelbäck, O. Hilborn, and P. Jerčić, "Evaluating classifiers for emotion recognition using EEG," in *Foundations of Augmented Cognition: 7th International Conference, AC 2013, Held as Part of HCI International 2013, Las Vegas, NV, USA, July 21–26, 2013. Proceedings 7*, pp. 492–501, Springer, 2013.

10. J. Chen, B. Hu, P. Moore, X. Zhang, and X. Ma, "Electroencephalogram-based emotion assessment system using ontology and data mining techniques," *Applied Soft Computing*, vol. 30, pp. 663–674, 2015.

11. A. M. Bhatti, M. Majid, S. M. Anwar, and B. Khan, "Human emotion recognition and analysis in response to audio music using brain signals," *Computers in Human Behavior*, vol. 65, pp. 267–275, 2016.

12. R.-N. Duan, J.-Y. Zhu, and B.-L. Lu, "Differential entropy feature for EEG-based emotion classification," in *2013 6th International IEEE/EMBS Conference on Neural Engineering (NER)*, pp. 81–84, IEEE, 2013.

13. N. Zhuang, Y. Zeng, K. Yang, C. Zhang, L. Tong, and B. Yan, "Investigating patterns for self-induced emotion recognition from EEG signals," *Sensors*, vol. 18, no. 3, p. 841, 2018.

14. M. Z. Soroush, K. Maghooli, S. K. Setarehdan, and A. M. Nasrabadi, "Emotion recognition using EEG phase space dynamics and poincare intersections," *Biomedical Signal Processing and Control*, vol. 59, p. 101918, 2020.

15. K. R. Scherer, "What are emotions? and how can they be measured?," *Social Science Information*, vol. 44, no. 4, pp. 695–729, 2005.

16. A. J. Gerber, J. Posner, D. Gorman, T. Colibazzi, S. Yu, Z. Wang, A. Kangarlu, H. Zhu, J. Russell, and B. S. Peterson, "An affective circumplex model of neural systems subserving valence, arousal, and cognitive overlay during the appraisal of emotional faces," *Neuropsychologia*, vol. 46, no. 8, pp. 2129–2139, 2008.

17. D. O. Bos et al., "EEG-based emotion recognition," *The Influence of Visual and Auditory Stimuli*, vol. 56, no. 3, pp. 1–17, 2006.

18. J. A. Russell, "Affective space is bipolar," *Journal of Personality and Social Psychology*, vol. 37, no. 3, p. 345, 1979.

19. W. Lin, C. Li, and S. Sun, "Deep convolutional neural network for emotion recognition using EEG and peripheral physiological signal," in *International Conference on Image and Graphics*, pp. 385–394, Springer, 2017.

20. S. Tripathi, S. Acharya, R. D. Sharma, S. Mittal, and S. Bhattacharya, "Using deep and convolutional neural networks for accurate emotion classification on deap dataset," in *Twenty-Ninth IAAI Conference*, 2017.

21. K. He, X. Zhang, S. Ren, and J. Sun, "Deep residual learning for image recognition," in *Proceedings of the IEEE Conference on Computer Vision and Pattern Recognition*, pp. 770–778, 2016.

22. A. Krizhevsky, I. Sutskever, and G. E. Hinton, "Imagenet classification with deep convolutional neural networks," *Advances in Neural Information Processing Systems*, vol. 25, 2012.

23. S. Ren, K. He, R. Girshick, and J. Sun, "Faster R-CNN: Towards real-time object detection with region proposal networks," *Advances in Neural Information Processing Systems*, vol. 28, 2015.

24. M. Danelljan, A. Robinson, F. Shahbaz Khan, and M. Felsberg, "Beyond correlation filters: Learning continuous convolution operators for visual tracking," in *European Conference on Computer Vision*, pp. 472–488, Springer, 2016.

25. M. N. Dar, M. U. Akram, R. Yuvaraj, S. G. Khawaja, and M. Murugappan, "EEG-based emotion charting for parkinson's disease patients using convolutional recurrent neural networks and cross dataset learning," *Computers in Biology and Medicine*, vol. 144, p. 105327, 2022.

26. M. Shehab, L. Abualigah, Q. Shambour, M. A. Abu-Hashem, M. K. Y. Shambour, A. I. Alsalibi, and A. H. Gandomi, "Machine learning in medical applications: A review of state-of-the-art methods," *Computers in Biology and Medicine*, vol. 145, p. 105458, 2022.

27. M.-P. Hosseini, A. Hosseini, and K. Ahi, "A review on machine learning for EEG signal processing in bioengineering," *IEEE Reviews in Biomedical Engineering*, vol. 14, pp. 204–218, 2020.

28. R. Alhalaseh and S. Alasasfeh, "Machine-learning-based emotion recognition system using EEG signals," *Computers*, vol. 9, no. 4, p. 95, 2020.

29. M. Moshfeghi, J. P. Bartaula, and A. T. Bedasso, "Emotion recognition from EEG signals using machine learning," in *Bachelor's Thesis, School of Engineering, Blekinge Institute of Technology*, Karlskrona, Sweden, March 2013.

30. V. M. Joshi and R. B. Ghongade, "EEG based emotion detection using fourth order spectral moment and deep learning," *Biomedical Signal Processing and Control*, vol. 68, p. 102755, 2021.

31. F. Demir, N. Sobahi, S. Siuly, and A. Sengur, "Exploring deep learning features for automatic classification of human emotion using EEG rhythms," *IEEE Sensors Journal*, vol. 21, no. 13, pp. 14923–14930, 2021.

32. G. Placidi, P. Di Giamberardino, A. Petracca, M. Spezialetti, and D. Iacoviello, "Classification of emotional signals from the deap dataset," in *NEUROTECHNIX*, pp. 15–21, 2016.

33. M. A. Rahman, M. F. Hossain, M. Hossain, and R. Ahmmed, "Employing pca and t-statistical approach for feature extraction and classification of emotion from multichannel EEG signal," *Egyptian Informatics Journal*, vol. 21, no. 1, pp. 23–35, 2020.

34. M. M. Rahman, A. K. Sarkar, M. A. Hossain, M. S. Hossain, M. R. Islam, M. B. Hossain, J. M. Quinn, and M. A. Moni, "Recognition of human emotions using EEG signals: A review," *Computers in Biology and Medicine*, vol. 136, p. 104696, 2021.

35. D. Maheshwari, S. K. Ghosh, R. Tripathy, M. Sharma, and U. R. Acharya, "Automated accurate emotion recognition system using rhythm-specific deep convolutional neural network technique with multi-channel EEG signals," *Computers in Biology and Medicine*, vol. 134, p. 104428, 2021.

36. H. J. Yoon and S. Y. Chung, "EEG-based emotion estimation using bayesian weighted-log-posterior function and perceptron convergence algorithm," *Computers in biology and medicine*, vol. 43, no. 12, pp. 2230–2237, 2013.

37. J. Li, H. Hua, Z. Xu, L. Shu, X. Xu, F. Kuang, and S. Wu, "Cross-subject EEG emotion recognition combined with connectivity features and meta-transfer learning," *Computers in Biology and Medicine*, vol. 145, p. 105519, 2022.

38. M. R. Islam, M. M. Islam, M. M. Rahman, C. Mondal, S. K. Singha, M. Ahmad, A. Awal, M. S. Islam, and M. A. Moni, "EEG channel correlation based model for emotion recognition," *Computers in Biology and Medicine*, vol. 136, p. 104757, 2021.

39. R. C. Dhingra and S. Ram Avtar Jaswal, "Emotion recognition based on EEG using deap dataset," *European Journal of Molecular & Clinical Medicine*, vol. 8, no. 3, pp. 3509–3517, 2021.

40. D. Acharya, R. Jain, S. S. Panigrahi, R. Sahni, S. Jain, S. P. Deshmukh, and A. Bhardwaj, "Multi-class emotion classification using EEG signals," in *International Advanced Computing Conference*, pp. 474–491, Springer, 2020.

41. S. M. Alarcao and M. J. Fonseca, "Emotions recognition using EEG signals: A survey," *IEEE Transactions on Affective Computing*, vol. 10, no. 3, pp. 374–393, 2017.

42. Y.-T. Matsuda, T. Fujimura, K. Katahira, M. Okada, K. Ueno, K. Cheng, and K. Okanoya, "The implicit processing of categorical and dimensional strategies: An fMRI study of facial emotion perception," *Frontiers in human neuroscience*, vol. 7, p. 551, 2013.

43. R. Plutchik, "In search of the basic emotions," *PsycCRITIQUES*, vol. 29, no. 6, 1984.

44. A. S. Cowen and D. Keltner, "Self-report captures 27 distinct categories of emotion bridged by continuous gradients," *Proceedings of the National Academy of Sciences*, vol. 114, no. 38, pp. E7900–E7909, 2017.

45. J. A. Russell, "A circumplex model of affect.," *Journal of Personality and Social Psychology*, vol. 39, no. 6, p. 1161, 1980.

46. S. Koelstra, C. Muhl, M. Soleymani, J.-S. Lee, A. Yazdani, T. Ebrahimi, T. Pun, A. Nijholt, and I. Patras, "Deap: A database for emotion analysis; using physiological signals," *IEEE Transactions on Affective Computing*, vol. 3, no. 1, pp. 18–31, 2011.

47. S. Katsigiannis and N. Ramzan, "Dreamer: A database for emotion recognition through EEG and ECG signals from wireless low-cost off-the-shelf devices," *IEEE Journal of Biomedical and Health Informatics*, vol. 22, no. 1, pp. 98–107, 2017.

48. P. Lang and Margaret M. Bradley. "The international affective picture system (IAPS) in the study of emotion and attention," *Handbook of Emotion Elicitation and Assessment*, vol. 29, pp. 70–73, 2017.

49. R. Subramanian, J. Wache, M. K. Abadi, R. L. Vieriu, S. Winkler, and N. Sebe, "Ascertain: Emotion and personality recognition using commercial sensors," *IEEE Transactions on Affective Computing*, vol. 9, no. 2, pp. 147–160, 2016.

50. M. Soleymani, J. Lichtenauer, T. Pun, and M. Pantic, "A multimodal database for affect recognition and implicit tagging," *IEEE Transactions on Affective Computing*, vol. 3, no. 1, pp. 42–55, 2011.

51. W.-L. Zheng and B.-L. Lu, "A multimodal approach to estimating vigilance using EEG and forehead EOG," *Journal of Neural Engineering*, vol. 14, no. 2, p. 026017, 2017.

52. J. A. Miranda-Correa, M. K. Abadi, N. Sebe, and I. Patras, "Amigos: A dataset for affect, personality and mood research on individuals and groups," *IEEE Transactions on Affective Computing*, vol. 12, no. 2, pp. 479–493, 2018.

53. B. Lu, M. Hui, and H. Yu-Xia, "The development of native chinese affective picture system–a pretest in 46 college students," *Chinese Mental Health Journal*, vol. 19, no. 11, pp. 719–722, 2005.

54. P. Lakhan, N. Banluesombatkul, V. Changniam, R. Dhithijaiyratn, P. Leelaarporn, E. Boonchieng, S. Hompoonsup, and T. Wilaiprasitporn, "Consumer grade brain sensing for emotion recognition," *IEEE Sensors Journal*, vol. 19, no. 21, pp. 9896–9907, 2019.

55. A. T. Sohaib, S. Qureshi, J. Hagelbäck, O. Hilborn, and P. Jerčić, "Evaluating classifiers for emotion recognition using EEG," in *Foundations of Augmented Cognition: 7th International Conference, AC 2013, Held as Part of HCI International 2013, Las Vegas, NV, USA, July 21–26, 2013. Proceedings 7*, pp. 492–501, Springer, 2013.

56. Y.-H. Liu, C.-T. Wu, Y.-H. Kao, and Y.-T. Chen, "Single-trial EEG-based emotion recognition using kernel eigen-emotion pattern and adaptive support vector machine," in *2013 35th Annual International Conference of the IEEE Engineering in Medicine and Biology Society (EMBC)*, pp. 4306–4309, IEEE, 2013.

57. A. Khosla, P. Khandnor, and T. Chand, "A comparative analysis of signal processing and classification methods for different applications based on EEG signals," *Biocybernetics and Biomedical Engineering*, vol. 40, no. 2, pp. 649–690, 2020.

58. R. Nawaz, K. H. Cheah, H. Nisar, and V. V. Yap, "Comparison of different feature extraction methods for EEG-based emotion recognition," *Biocybernetics and Biomedical Engineering*, vol. 40, no. 3, pp. 910–926, 2020.

59. S. Abe and S. Abe, "Feature selection and extraction," *Support Vector Machines for Pattern Classification*, pp. 331–341, 2010.

60. R. Zebari, A. Abdulazeez, D. Zeebaree, D. Zebari, and J. Saeed, "A comprehensive review of dimensionality reduction techniques for feature selection and feature extraction," *Journal of Applied Science and Technology Trends*, vol. 1, no. 2, pp. 56–70, 2020.

61. S. Siuly, Y. Li, and Y. Zhang, "EEG signal analysis and classification," *IEEE Transactions on Neural Systems and Rehabilitation Engineering*, vol. 11, pp. 141–144, 2016.

62. M. Savadkoohi, T. Oladunni, and L. Thompson, "A machine learning approach to epileptic seizure prediction using electroencephalogram (EEG) signal," *Biocybernetics and Biomedical Engineering*, vol. 40, no. 3, pp. 1328–1341, 2020.

63. M. S. Hossain, S. U. Amin, M. Alsulaiman, and G. Muhammad, "Applying deep learning for epilepsy seizure detection and brain mapping visualization," *ACM Transactions on Multimedia Computing, Communications, and Applications (TOMM)*, vol. 15, no. 1s, pp. 1–17, 2019.

64. L. Stanković, M. Daković, and E. Sejdić, "Introduction to graph signal processing," *Vertex-Frequency Analysis of Graph Signals*, pp. 3–108, 2019.

65. G. Sharma, A. Parashar, and A. M. Joshi, "Dephnn: a novel hybrid neural network for electroencephalogram (EEG)-based screening of depression," *Biomedical Signal Processing and Control*, vol. 66, p. 102393, 2021.

66. H. Padole, S. D. Joshi, and T. K. Gandhi, "Early detection of alzheimer's disease using graph signal processing on neuroimaging data," in *2018 2nd European Conference on Electrical Engineering and Computer Science (EECS)*, pp. 302–306, IEEE, 2018.

67. M. Defferrard, X. Bresson, and P. Vandergheynst, "Convolutional neural networks on graphs with fast localized spectral filtering," *Advances in Neural Information Processing Systems*, vol. 29, 2016.

68. S. Jang, S.-E. Moon, and J.-S. Lee, "EEG-based video identification using graph signal modeling and graph convolutional neural network," in *2018 IEEE International Conference on Acoustics, Speech and Signal Processing (ICASSP)*, pp. 3066–3070, IEEE, 2018.

69. T. Song, W. Zheng, P. Song, and Z. Cui, "EEG emotion recognition using dynamical graph convolutional neural networks," *IEEE Transactions on Affective Computing*, vol. 11, no. 3, pp. 532–541, 2018.

70. W.-L. Zheng, J.-Y. Zhu, and B.-L. Lu, "Identifying stable patterns over time for emotion recognition from EEG," *IEEE Transactions on Affective Computing*, vol. 10, no. 3, pp. 417–429, 2017.

71. M. Li, H. Xu, X. Liu, and S. Lu, "Emotion recognition from multichannel EEG signals using k-nearest neighbor classification," *Technology and Health Care*, vol. 26, no. S1, pp. 509–519, 2018.

72. S. Taran and V. Bajaj, "Emotion recognition from single-channel EEG signals using a two-stage correlation and instantaneous frequency-based filtering method," *Computer Methods and Programs in Biomedicine*, vol. 173, pp. 157–165, 2019.

73. Y. Wang, S. Qiu, C. Zhao, W. Yang, J. Li, X. Ma, and H. He, "EEG-based emotion recognition with prototype-based data representation," in *2019 41st Annual International Conference of the IEEE Engineering in Medicine and Biology Society (EMBC)*, pp. 684–689, IEEE, 2019.

74. H. Cui, A. Liu, X. Zhang, X. Chen, K. Wang, and X. Chen, "EEG-based emotion recognition using an end-to-end regional-asymmetric convolutional neural network," *Knowledge-Based Systems*, vol. 205, p. 106243, 2020.

75. Y. Zhang, C. Cheng, and Y. Zhang, "Multimodal emotion recognition using a hierarchical fusion convolutional neural network," *IEEE Access*, vol. 9, pp. 7943–7951, 2021.

76. S. K. Khare and V. Bajaj, "An evolutionary optimized variational mode decomposition for emotion recognition," *IEEE Sensors Journal*, vol. 21, no. 2, pp. 2035–2042, 2020.

77. A. Bhattacharyya, R. K. Tripathy, L. Garg, and R. B. Pachori, "A novel multivariate-multiscale approach for computing EEG spectral and temporal complexity for human emotion recognition," *IEEE Sensors Journal*, vol. 21, no. 3, pp. 3579–3591, 2020.

8 Automated Detection of Alzheimer's Disease Using EEG Signal Processing and Machine Learning

Mahbuba Ferdowsi, Haipeng Liu,
Ban-Hoe Kwan, and Choon-Hian Goh

8.1 ALZHEIMER'S DISEASE AND CHALLENGES IN ITS EARLY DETECTION

Dementia's most prevalent form is Alzheimer's disease (AD) (Pirrone et al., 2022). With a predicted prevalence of 682 million by 2050, AD is to be the primary cause of dementia in several nations (Hogg and Watt, 2012). As a hereditary and sporadic neurodegenerative disease, AD is a prevalent cause of cognitive impairment in mid-life and later in life (Knopman et al., 2021). An amnestic cognitive impairment characterizes AD's prototypical appearance, while nonamnestic cognitive impairment characterizes its less common variations (Knopman et al., 2021). Common symptoms of AD include memory loss and impairment in at least one domain of cognition such as calculation, praxis, gnosis, executive functions, or language (Hogg and Watt, 2012). Other neurodegenerative and cerebrovascular disorders (e.g., moyamoya disease, intracranial atherosclerosis) can also cause cognitive impairment and be the confounders of AD (Zhou et al., 2015; Li et al., 2022; Wang et al., 2023).

Although the pathology of AD is not fully understood, some risk factors have been identified. The most significant and immutable risk factor for practically all neurodegenerative disorders, including AD, is aging. As people age, the human brain goes through a number of genetic, biochemical, anatomical, and functional changes that may increase our vulnerability to cognitive illnesses such as neurodegeneration and dementia, including AD (Trollor and Valenzuela, 2001). Due to an excess of all-cause mortality in men over the age of 45, dementia affects more women than men in the global elderly population, for instance, among those over 65 (Wu et al., 2017). Mild cognitive impairment (MCI) is frequently defined by a loss of cognition, such as remembering or thinking, that is greater than what is usual for the age (Amezquita-Sanchez et al., 2019). A high chance of developing AD or other neurological diseases exists in MCI patients (McBride et al., 2015). The prevalence of MCI may also be

DOI: 10.1201/9781003479970-8

higher in men than in women, according to one study that revealed an OR of 1.54 (95% CI 1.2–2.0) for MCI in men than women in a group of European Americans (Petersen, 2010). Other studies have found no difference between the sexes in the prevalence of dementia, or that it is higher in women (Petersen, 2010).

The diagnosis of AD depends on the patient's medical history, family history, and behavioral observations. Considering the multiple risk factors and confounders of AD, accurate early detection of AD plays a key role in improving the efficiency of clinical intervention. Positron emission tomography (PET), cerebrospinal fluid, and magnetic resonance imaging (MRI) have been proposed for the diagnosis of AD, but they are inappropriate for clinical practice and daily monitoring due to high expense or invasiveness (Rossini et al., 2020; Al-Nuaimi et al., 2021). Despite its importance, the early diagnosis of AD is difficult due to the lack of clear diagnostic standards, the confounding effect of other neurological disease, and insufficient understanding of its pathophysiology.

8.2   ELECTROENCEPHALOGRAM AND ITS POTENTIAL FOR AD DETECTION

By measuring postsynaptic potentials on scalp electrodes, EEG records electrical activity in the cerebral cortex from 1,000 identically oriented neurons, which may contain AD-relevant electrophysiological features (Nobukawa et al., 2020). The spatial revolution of EEG depends on the quantity and arrangement of scalp electrodes. Most clinical and research applications specify the electrode locations and names using the International 10–20 system (Nobukawa et al., 2020). This strategy ensures that electrode names are uniform across laboratories. Most clinical applications make use of 19 recording electrodes (Table 8.1). Typically, the EEG signals are preprocessed to eliminate undesirable noise and artifacts and to highlight the frequency bands of interest. To explore the cerebral activity associated with various cognitive and behavioral processes, the EEG signals can be studied using a variety of signal processing techniques, such as discrete wavelet transform (DWT), Burg method, and the average periodogram (AP) (Robert et al., 2020).

The temporal dynamics of brain activity and how it varies over time can be revealed by EEG, enabling the multimodal observation of various pathophysiological phenomena. Overall, EEG technology provides a noninvasive and affordable solution for the observation of AD-associated brain activity, as well as the development of new approaches for the diagnosis, treatment, and management of AD (Biagetti et al., 2021). Those who have AD in particular frequently display aberrant EEG patterns, such as diminished power and coherence in particular frequency bands, as well as increased delta and theta activity. These alterations in EEG activity are a reflection of the underlying neurophysiological abnormalities brought on by AD, including neuronal loss, synaptic dysfunction, and aberrant network connections (Cassani and Falk, 2020). EEG also discloses the temporal dynamics of brain activity, which enables researchers to investigate the development of AD by observing the evolution of relevant EEG features.

TABLE 8.1

Summary of Artificial Intelligence Models for Alzheimer's Disease Detection Using EEG

Authors	Sample size	No. of electrode channels	Preprocessing methods	Models	Performance
(Thakare and Pawar, 2016)	15 (N/A)	19	Wavelet transform	Support vector machine	Accuracy = 0.95
(Trambaiolli et al., 2017)	34 (22 AD and 12 HC)	19	NA	Support vector machine	Accuracy = 0.91
(Houmani et al., 2018)	169 (22 SCI, 98 MCI and 49 AD)	30	Epoch-based entropy	Support vector machine	Accuracy = 0.92, Specificity = 1, Sensitivity = 0.88
(Fiscon et al., 2018)	109 (86 AD and 23 HC)	19	Spectral analysis (Wavelet and Fourier)	Decision tree	The accuracy rates for HC vs AD, HC vs MCI, and MCI vs AD was 0.83, 0.92, and 0.79, respectively
(Vecchio et al., 2020)	295 (175 AD and 120 HC)	19 or 32 (multi-centre study)	(N/A)	Support vector machine	AUC = 0.97
(Nobukawa et al., 2020)	34 (16 AD and 18 HC)	18	(N/A)	Support vector machine	AUC = 1, Accuracy = 1
(Cicalese et al., 2020)	8 HC, 8 MCI, 6 mild AD, and 7 moderate/severe AD	32	(N/A)	Linear discriminant analysis	Accuracy = 0.79
(Oltu et al., 2021)	35 (16 MCI, 8 AD, and 11 HC)	19	Discrete wavelet transform, power spectral density and coherence	Bagged tree (Bagging)	Accuracy = 0.97, Sensitivity = 0.96, Specificity = 0.98
(Amini et al., 2021)	192 (64 AD, MCI, and HC)	19	Time-dependent power spectrum descriptors	Convolutional neural networks	Accuracy = 0.82, ROC(AUC) = 0.99
(Chedid et al., 2022)	43 (20 AD and 23 HC)	32	Band-wise power spectral density	Logistic regression	Accuracy = 0.81
(Alsharabi et al., 2022)	88 (35 HC, 31 mild AD and 22 moderate AD)	22	Discrete wavelet transform	K-nearest neighbor	Accuracy = 1, ROC (AUC) = 1

Notes: ROC, receiver operating characteristic curve; AUC, area under curve; N/A, not available; HC, healthy control; MCI, mild cognitive impairment; SCI: subjective cognitive impairment; AD, Alzheimer's disease.

8.3 ARTIFICIAL-INTELLIGENCE-ASSISTED EEG ANALYSIS: A PROMISING APPROACH FOR THE EARLY DETECTION OF AD

Some artificial intelligence (AI) techniques including support vector machine (SVM), k-nearest neighbor (KNN), random forest (RF), decision tree, logistic regression (LR), linear discriminant analysis (LDA), AdaBoost, bagging, artificial neural networks (ANN), convolutional neural networks (CNNs) and spatial temporal convolutional networks (STCNs) have been applied to categorize EEG features and identify subjects with high risks of AD. Traditional diagnostic methods rely on mass-univariate statistics where the regions are assumed to work independently (Subasi, 2020). In contrast, AI techniques specifically leverage the correlation among the areas. A further difference between mass univariate statistical methods and AI techniques is that the latter use differences at the group level while the former use implications at the subject level (Subasi, 2020). Thus, AI-assisted analysis improves the reliability and computational capacity of EEG signal analysis. Currently, there is a scarcity of comprehensive comparisons of different AI-assisted EEG analysis methods for early detection of AD. To address this gap, we systematically reviewed recent studies.

8.4 METHODS: LITERATURE SEARCH STRATEGY

The literature search strategy investigates AD from EEG signals utilizing various signal processing techniques and AI. As part of the literature review methodology design, we first discussed the main topics of research and identified keywords and phrases. First, we looked for "Alzheimer", "EEG", and either "detection" or "classification" together. Next, the Boolean operator "and" was used to link the keywords "Artificial intelligence" to both "detection" and "classification". Literature searches were conducted in numerous databases: PubMed (https://pubmed.ncbi.nlm.nih.gov/), Scopus (https://pubmed.ncbi.nlm.nih.gov/), MDPI (https://www.mdpi.com/), BMC (https://www.biomedcentral.com/), ScienceDirect (https://www.sciencedirect.com/), IEEE Xplore (https://ieeexplore.ieee.org/Xplore/home.jsp), BMJ Open (https://bmjopenquality.bmj.com/), Frontiers (https://www.frontiersin.org/), and PLoS One (https://journals.plos.org/plosone/) as our key information sources, as well as Google Scholar (https://scholar.google.com/) for double check. In order to narrow our search further, we used the databases' filters based on date, author, type (i.e., journal article, conference paper, thesis, etc.), and other criteria. With the exception of six publications, every paper that was looked for was published between 2015 and 2023, and all were written in English. The papers that were unrelated to using AI and signal processing to detect Alzheimer's disease were removed during full-text reading. The finally selected papers were analyzed in details to assess the effectiveness of AI-based Alzheimer's disease detection approaches. The flowchart of the review method is shown in Figure 8.1.

8.4.1 INFORMATION RETRIEVAL FROM SCIENTIFIC DATABASES

This section primarily covers the scientific research, exploring and accumulating from the pertinent sources. Over 100 research papers were initially retrieved based on their titles and abstracts using keywords and Boolean operators (AND, OR, and NOT).

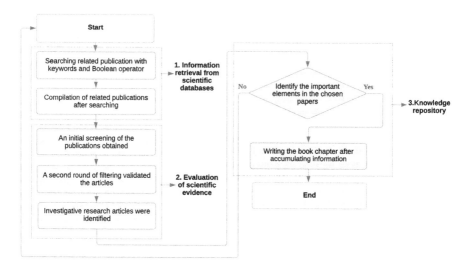

FIGURE 8.1   Flowchart for literature search methodology.

8.4.2   Evaluation of Scientific Evidence

We evaluated the publications we collected regarding their scientific evidence. First, we excluded the items that were irrelevant or only tangentially related to AD detection during the abstract reading, which left about 70 items for scrutiny. Next, we read the full texts of those articles and selected only the papers focusing on AD detection with EEG signal and AI. We were left with 43 papers for analysis.

8.4.3   Knowledge Repository

We finally analyzed the selected papers in detail to extract essential components including the datasets and methodological details, as well as the benefits, drawbacks, and potential enhancements of the relevant stakeholders. If limited details could be extracted, we fine-tuned the keywords to restart another iteration of the literature search.

8.5   FRAMEWORK OF AD DETECTION

Here, we cover the framework of AD detection, including EEG waves, data preprocessing, signal processing, and the role of AI.

8.5.1   EEG Waves

EEG waves are electrical signals produced by the electrical activity of the brain. Depending on their frequency and amplitude properties, EEG waves can be categorized into several frequency bands or wave types including delta waves (0.5–4 Hz), theta waves (4–8 Hz), alpha waves (8–12 Hz), beta waves (12–30 Hz) and

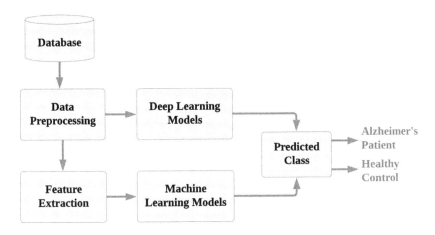

FIGURE 8.2   General flowchart for AD detection.

gamma waves (30–100 Hz). Delta waves, the slowest EEG wave type, are usually associated with unconsciousness and deep sleep. Theta waves are a common feature of sleepiness, daydreaming, and various forms of meditation. It is common to find alpha waves when the brain is alert but calm, such as when people close their eyes and unwind. Beta waves are associated with active thought, concentration, and attention. Gamma waves, the fastest subset of the EEG, are associated with higher-order cognitive processes such as perception, consciousness, and attention.

The electrode placement on the scalp can also influence the recorded EEG signals. Except when high-density arrays are employed, the International 10–20 system specifies the locations and names of the electrodes in the majority of clinical and research applications (Nobukawa et al., 2020). This approach guarantees that electrode names are consistent among laboratories. The majority of clinical applications employ 19 recording electrodes (see Table 8.1).

The raw EEG signals often contain various environmental and physiological noises (e.g., powerline noise and motion artefact). Therefore, preprocessing is essential to remove noises and unwanted components. The preprocessed EEG signals can be input into the deep learning algorithm for automatic classification, or be further processed to extract some dementia-associated features for the classification based on machine learning (Figure 8.2).

8.5.2 Data Preprocessing

Artifacts in the EEG signal being monitored are inevitable and can interfere with the analysis. Any signal that is measured but does not originate from the brain is considered an artifact. The cause of artifacts could be from the problems with the device, such as defective electrodes, line noise, high electrode impedance, or confounding physiological activities. Physiological artifacts, e.g., eye blinks, movement, heart activity, and muscular activity, are constant and difficult to eliminate during signal recording, and they can imitate cognitive activities, affecting the detection of AD-related issues.

To reject components with a 50/60 Hz frequency, several filtering techniques are used, such as a bandpass infinite impulse response elliptic digital filter (Alsharabi et al., 2022). Some artifacts overlap with EEG waves in frequency bands, where basic filters could not effectively denoise the signals. Therefore, blind source separation algorithms have been proposed to remove artifacts from EEG signals (Alsuradi et al., 2020). These include principal component analysis and independent component analysis, and several of these algorithms have proved successful at removing the majority of physiological artifacts (Alsuradi et al., 2020).

8.5.3 Signal Processing and Feature Extraction

Several signals processing techniques, including as time-frequency analysis, coherence analysis, and event-related potential analysis, can be used to detect AD-related EEG features of brain activities associated with various cognitive and behavioural processes.

8.5.3.1 Discrete Wavelet Transforms

A method for examining the local properties of a nonstationary signal in the temporal and spectral domains is the wavelet transform (Subha et al., 2010). The wavelet transform's key benefit is that it offers the best time-frequency resolution by using narrow window sizes for high frequencies and vast window sizes for low frequencies, which suits the wide frequency bands of EEG signals. Continuous wavelet transform and discrete wavelet transform (DWT) are the two different types of wavelet transform. In DWT, a scaling and a wavelet function connected to low-pass and high-pass filters is used to evaluate the signal and divide it into a number of sub frequency bands (Oltu et al., 2021). The approximation coefficients are the outputs of these filters. Sub-band decomposition, a filtering technique, is used to produce the low-frequency bands. Recursive sub-band decomposition can be carried out up until the necessary frequency ranges are attained or there is no more room for sub-sampling. The Nyquist theorem states that at each level, the frequency resolutions double as the time resolution decreases. Thus, it is possible to attain high time resolution at high frequencies as well as high frequency resolution at low frequencies (Subha et al., 2010).

8.5.3.2 Power Spectrum Density

Both the Burg method and the average periodogram (AP) approach can be used to estimate power spectrum density (PSD) of EEG signals (Oltu et al., 2021; Chedid et al., 2022). With the AP method, to estimate the PSD, the signal is split into segments. The periodogram, which describes how a signal's energy is distributed among its frequency components, is computed for each segment. The average of the periodograms is then calculated. The magnitude of each frequency component is computed based on Fourier transform. The prominent frequencies and their relative amplitudes can be derived from the periodogram. Segmenting a nonstationary signal can aid in capturing its underlying spectral properties that are time-varying. The resultant

PSD derived from different segments can offer a more precise representation of the spectral content of the signal.

Segment length selection and overlap have an effect on the precision of the anticipated PSD. Longer segments typically offer better frequency resolution but worse time resolution, while shorter segments typically give better time resolution but worse frequency resolution. A trade-off between these factors needs to be made when deciding on segment length and overlap. The autoregressive (AR) model, on which the Burg approach is based, models the signal as a linear combination of its prior values, with the model's coefficients calculated from the data (Wang et al., 2015). The AR model coefficients can then be used to estimate the PSD of the signal (Li and Wei, 2020).

8.5.3.3 Extracted Features

Coherence quantifies the degree of synchrony between various brain regions and indicates the functional connectivity of neural networks. The posterior regions of the brain show decreased coherence in Alzheimer's patients (He et al., 2007). PSD describes the distribution of power in the frequency bands of the EEG signal (delta, theta, alpha, beta, and gamma). When someone has Alzheimer's disease, power diminish can be observes in alpha and beta bands, while increased power appears in delta and theta bands.

An EEG signal's fractal dimension is a measure of its complexity and irregularity that considers the underlying brain networks (Goh et al., 2009). Alzheimer's patients have been found to have a reduced fractal dimension (Woyshville and Calabrese, 1994). Entropy is a measure of the EEG signal's randomness and unpredictability, and it might indicate the complexity of neural activities. Logarithmic band power, standard deviation, variance, kurtosis, average energy, root mean square, and norm are also proposed as features derived from the EEG signal (Alsharabi et al., 2022).

8.5.4 Applications of AI in AD Detection

AD is characterized by extensive, progressive neurodegenerative changes, making diagnosis difficult. AD cannot be accurately diagnosed with a single test or biomarker. Instead, the diagnosis relies on a combination of heterogeneous features derived from multimodal data (e.g., medical history, lab test, clinical imaging, etc.), where AI could ease the diagnosis and improve its accuracy based on prior knowledge. The analysis of EEG data for the diagnosis of AD using AI algorithms and other cutting-edge computational methods is gaining interest (Pandya et al., 2020). These methods could lead to earlier and more precise disease diagnoses by assisting in the detection of tiny changes in EEG patterns.

A number of ways exist by which AI can aid in identifying AD at an early stage (Counts et al., 2017; De Roeck et al., 2019). AI systems can analyze large, multimodal clinical databases to discover patterns and traits associated with AD by detecting small changes in brain structure and function that might be signs of AD in its early stages (Pegueroles et al., 2016). As a result, patients may receive more accurate diagnoses and better treatment. Genetic information and other biomarkers can be analyzed by AI to create individualized treatment programs for patients based on their unique traits (Wisniewski, 2017). AI could identify possible drug targets, enabling individualized treatment of AD based on biological data. AI can be used to create tools and programs that help caregivers optimize their services to improve the living quality of people with AD. Next we discuss some details on the algorithms used in recent studies on the AI-based diagnosis of AD.

8.5.4.1 Support Vector Machine

Support vector machine (SVM) seeks out a classification function that, to the best of its ability, distinguishes between instances of the two classes (Irankhah, 2020). If the dataset can be separated into two classes in a linear fashion, a hyperplane function that does this is a classification function that passes through the intersection of the two classes. The best function can be identified by maximizing the difference between the two classes. The hyperplane's distance between any two classes is known as the margin. It is defined as the shortest path between two instances on the hyperplane that are close to one another. This formulation enables the choice of the best hyperplanes as SVM solutions and margin optimization.

8.5.4.2 K-nearest Neighbors

K-nearest neighbor (KNN) works under the premise that the training set and the sample are compared using various distance measurements. To classify a new instance, the algorithm identifies the k training instances that are most similar to it and uses their classifications as a basis for determining the classification of the new instance (Irankhah, 2020). According to a majority vote among the training data classes that are closest to the new instance, a label is issued.

8.5.4.3 Random Forest

When building a decision tree using the random forest (RF) classifier, the algorithm randomly selects samples from a pool of N samples, which is the same size as the original training set, with replacement. In other words, it chooses samples from the pool randomly and allows the same sample to be selected more than once. The learning algorithm creates a classifier from a sample, and the final classifier is produced by merging all of the classifiers produced from the numerous trials. Every classifier casts a vote for the class that an instance belongs to, and the instance is then categorized as a member of the class that received the most votes. If more than one class receives the most votes overall, a random selection is made to determine the winner. A bootstrap replica of the input data that is independently created serves as the foundation for each tree in the ensemble (Zhao et al., 2021).

8.5.4.4 Decision Tree

A decision tree is made up of several divide-and-conquer-based decision tests with a tree-structure (Baglat et al., 2020). A decision tree consists of leaf nodes and branches. The root node of the tree is at the top, and nodes are used to denote the feature tests (split) that are used to divide the data. The leaf nodes reflect the label of data, and the branches show the routes that should be taken in light of the test findings. This method typically recurses during the categorization process. According to the chosen split, the given data is divided into subsets in each step, and each subset is used as the given data in the following phase.

8.5.4.5 Logistic Regression

Logistic regression (LR) is a supervised machine learning approach often utilized in binary classification (Zhao et al., 2021). Logistic function predicts the likelihood that the input belongs to the positive class. The logistic function, also called the sigmoid function, converts any real-valued input to a range (0,1). The LR model learns the coefficient values by minimizing a loss function that gauges the discrepancy between the true labels of the training examples and the anticipated probabilities. The LR model then forecast the likelihood that a new input would fall into the positive category based on a threshold which is commonly 0.5.

8.5.4.6 Linear Discriminant Analysis

A method for reducing dimensions used frequently for supervised classification issues is linear discriminant analysis (LDA) (Tavares et al., 2019). LDA aims to minimize class overlap and maximize class separability by projecting the dataset onto a lower-dimensional space. LDA is a supervised technique that employs both the input data and the class labels to find linear discriminants that optimize the separation between various classes. It can greatly lower the cost of computing and can also be used in data preprocessing to reduce the number of features.

8.5.4.7 AdaBoost

AdaBoost is an ensemble ML algorithm for classification and regression tasks. In classification, it can enhance the accuracy of weak classifiers, i.e., those that perform marginally better than random guessing, by combining them into a more powerful classifier using iterative training on various subsets. A weighted mixture of all the weak classifiers produces the final classifier based on their performance. The ability of AdaBoost to manage high-dimensional data with numerous attributes is one of its advantages. It can also deal with class imbalance, which occurs when one class has a disproportionately large number of instances compared with another. Overfitting, however, might result from AdaBoost's sensitivity to noisy input and outliers (Tavares et al., 2019).

8.5.4.8 Bagging

Bagging, sometimes referred to as bootstrap aggregating, is an ensemble learning method used for classification and regression applications (Oltu et al., 2021). By building numerous models using various subsets of the training data, it is possible to increase the accuracy of models that are prone to overfitting, such as decision trees. These models are then integrated to provide a final classifier that offers a more reliable and precise prediction. Bagging can handle complex data with noise and has a number of advantages, such as lowering variance and overfitting. However, for the models that are already stable with little volatility, bagging barely brings improvement.

8.5.4.9 Artificial Neural Networks

Artificial neural networks (ANN) were developed to simulate how the human brain processes information and learns from experience to predict future outcomes and identify patterns. This is achieved through a computer program that applies ML and pattern recognition algorithms to create effective predictive models. To build an ANN, the analyst needs to determine the appropriate number of nodes for the network, how they should be connected, the optimal weight values for each connection, and how the training should be carried out. These decisions are crucial in building an effective ANN (Kumar Shrivastava and Rajak, 2020).

In an ANN, input data is received through input nodes and multiplied by weight values that are stored in connections. The output of the previous layer is then used as input for the next layer, which is added as a hidden layer between the input and output layers. This multilayer approach is commonly used in neural networks, where a transfer function is applied in the hidden layer to generate the resulting value.

The output node then receives the prediction value, which is based on the input data. The input and output layers are composed of nodes, with the input layer consisting of input variables and the output layer consisting of the target variable. In neural networks, each input variable with its own unit is called a unit. The model can be trained though both forward and backward propagation. However, a drawback of ANN is that the resulting model can be complex.

8.5.4.10 Convolutional Neural Networks

Convolutional neural networks (CNNs) were designed to handle data with a grid-like structure, such as photos and videos. Therefore, the CNNs consist of multiple layers that each perform a specific task on the input data. Convolutional layer, pooling layers, and fully connected layers are examples of these layers. During the convolutional layer, the network applies filters to the input image that can recognize various visual patterns, such as edges, corners, and textures. These filters are developed through training.

The pooling layer then reduces the spatial dimensions of the feature maps, making the network less computationally complex and more adaptable to changes in the input. The completely linked layer, also known as the thick layer, is responsible for making the final classification determination using the features extracted from the previous layers (Lopes et al., 2023). The output of the final pooling layer is flattened into a one-dimensional vector and passed through one or more fully linked layers (Figure 8.3).

The final classification probabilities are obtained by running the output of the final fully connected layer through an activation function like SoftMax. CNNs have successfully completed many different image and video recognition tasks, such as object

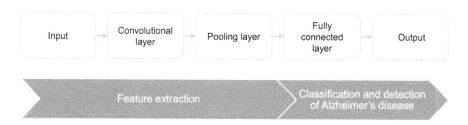

FIGURE 8.3   General overview of convolutional neural networks.

detection, face recognition, and scene segmentation. They are extensively used in both industry and academia and are continually being improved for better performance and generalization through changes in their architecture and training methods.

8.5.4.11 Spatial Temporal Convolutional Network

Spatial temporal convolutional networks (STCNs) are proposed for handling and evaluating sequential data, such time-series data or video. They are based on the idea of a feedforward CNN. By introducing temporal information into the model, STCNs expand the capabilities of conventional CNN design by enabling them to recognize and analyze the transient changes in input data.

In STCNs, spatial and temporal features are extracted from input data using 3D convolutional layers; the kernel size is commonly a 3D tensor (height, width, time), where height and width denote the spatial dimensions and time denotes the temporal dimension. The 3D tensor is dragged over the input data to conduct the convolution procedure. A significant hyperparameter in STCNs is kernel size, which must be carefully tuned to achieve optimal performance. STCNs can improve the efficiency and accuracy of extracting spatial and temporal information from EEG data (You et al., 2020).

8.6 DISCUSSION: COMPARISON OF STATE-OF-THE-ART TECHNIQUES

As summarized in Table 8.2, the technical innovations concentrate on the classification and EEG feature extraction. Traditional EEG decomposition is still widely

TABLE 8.2
Comparison of State-of-the-Art Techniques

Authors	Methods	Type of EEG decomposition	EEG features	Benefits	Limitations
(Cejnek et al., 2021)	1. Linear neural unit with Gradient Descent adaptation. 2. Class: (AD+MCI) as positive and HC as negative	Traditional	(N/A)	A customized approach was developed to classify AD.	Small sample size (only seven MCI patients)
(Miltiadous et al., 2021)	1. Butterworth band-pass filter. 2. DT and RF with 10-fold CV. 3. Class: AD, HC, and frontotemporal dementia	Adaptive	Mean, variance, IQR, and frequency band energy	The WEKA data platform was applied for standardized classification.	Small sample size (28 participants)

(Continued)

TABLE 8.2 (*Continued*)
Comparison of State-of-the-Art Techniques

Authors	Methods	Type of EEG decomposition	EEG features	Benefits	Limitations
(Safi et al., 2021)	1. Filtering into brain frequency bands, DWT, and empirical mode decomposition 2. KNN 3. Class: AD, mild AD and HC	Traditional	Variance, kurtosis, skewness, Shannon entropy, sure entropy, and Hjorth parameters	Signals with nonstationary properties can be decomposed empirically using discrete wavelet transforms.	Empirical mode decompositi-on is computation-ally expensive and can take a long time to process.
(Fakhri et al., 2023)	1. Variational autoencoder 2. Welch method 3. SVM 4. Class: AD, MCI, and HC	Adaptive	Average PSD, variance, mean, and zero-crossing rate	Data augmentation empowers transferring knowledge between domains because it produces new variants of the input data that are more representative of the target domain.	Variational autoencoder may cause overfitting and have an impact on the model's generalizability.
(Puri et al., 2023)	1. Low-complexity orthogonal wavelet filter banks with vanishing moments. 2. SVM with 10-fold CV. 3. Class: AD and HC.	Traditional	Higuchi's fractal dimension and Katz's fractal dimension	1. Smooth wavelet function enables algorithms to precisely capture the signal's high-frequency features. 2. This allows the analysis of signals with a wide range of frequency content at multiresolution.	Smooth wavelet function has only a few filter coefficients, which restricts its applicability to different types of EEG signals.

Notes: N/A; not available, SVM, support vector machine; CV, cross validation; KNN, k nearest neighbour; DT, decision tree; RF, random forest; PSD, power spectral density; HC, healthy control; MCI, mild cognitive impairment; AD, Alzheimer's disease.

used, which sometimes involves manual identification, making it time-consuming and prone to human error. On the other hand, adaptive EEG decomposition entails running an algorithm on the EEG signal to disentangle the required brainwave activity from any unwanted noise or artifacts. This is accomplished by combining time–frequency analysis, advanced signal processing, and ML algorithms. Next we compare the benefits and drawbacks of the current cutting-edge approaches.

8.7 BENEFITS OF AI IN EEG-BASED AD DIAGNOSIS

The application of AI-empowered EEG analysis greatly improves AD detection. The latest algorithms enable the detection of AD-associated minor changes in brain activities by advanced signal processing in multiple domains and fine-grained classification. These advances provide higher accuracy and sensitivity in detecting early symptoms of AD than manual operation since the AD-associated changes in EEG are difficult to identify in visual inspection.

An AI system enables the swift and accurate processing of large volumes of EEG data, providing noninvasively unique reference for personalized diagnosis and treatment of AD. By reducing the potential for subjective bias in human interpretation, it can offer an objective examination of EEG data. Compared with other diagnostic techniques like MRI or PET scans, it is a more affordable option, without the need for any new costly and time-consuming examinations or lab tests.

AI algorithms can pinpoint specific regions of the brain affected by AD, which provides guidance to improve the treatment efficacy of medication. Furthermore, AI enables the integrated analysis of multimodal big AD datasets including electronic health records, neuroimaging data, genetic data, and behavioral data to increase the validity and precision of AD diagnosis, prognosis, and therapy monitoring (Termine et al., 2021; Batko and Ślęzak, 2022). It is possible to disclose more underlying pathophysiological mechanisms of AD and hasten the development of individualized treatment and management. In order to help physicians and caregivers make better decisions, AI systems can offer real-time decision support. Through streamlining procedures, easing the load on physicians, and automating repetitive operations, AI can improve clinical workflows. To summarize, AI is playing an increasingly important role in improving the diagnosis, treatment, and management of AD.

8.8 LIMITATIONS

In comparison with existing methods (e.g., neuropsychological assessment, genetic testing) for early detection of AD, AI-based detection techniques are still under development and need further validation (Weston et al., 2017; Loewenstein et al., 2018). Despite the improved sensitivity in detecting early AD symptoms, many symptoms are subtle and easily misinterpreted as general signs of aging or other illnesses (Zhang et al., 2019; Fujita et al., 2022). The improved specificity and accuracy of AD detection, as well as fine-grained classification of dementia, are the challenges to the AI systems (Kelly et al., 2019). The ethical ramifications of early AD identification may include issues with informed consent, privacy, and possible effects on mental health and wellbeing.

AI-based AD detection methods, especially the deep learning algorithms, entail processing and analyzing big datasets, which can be technologically demanding and computationally burdensome. Deep learning models have many variables that need to be tuned throughout training, and they also demand high computing power for inference—the process of making predictions on fresh data. The dependence on high processing capacity and computational resources limits the real-time applications.

8.9 CONCLUSION AND FUTURE WORKS

The use of AI has high potential for significant advancements in neurological illnesses and has shown good performance in the identification of AD. Future work can be considered to improve the state of the art from different aspects. First, to overcome the limitations on sample size that have been common in recent studies, big multimodal datasets are necessary to improve the applicability of the AI algorithms among different cohorts. Multicenter collaborations and data-sharing are encouraged to achieve this, where data is collected using the same recording and scanning protocols across sites.

In addition, more biomarkers can be considered to find AD-related features in different domains. Finally, advanced signal processing methods can be explored to denoise the EEG signal in different application scenarios. The integration of different AI approaches could also be considered as a major approach towards higher detection accuracy. With these improvements, AI may pave the way for notable advancements in early detection, individualized treatment, and effective management of AD in the years to come.

REFERENCES

Al-Nuaimi, A. H. et al. (2021) 'Robust EEG-based biomarkers to detect alzheimer's disease', *Brain Sciences*, 11(8), pp. 1–31. doi: 10.3390/brainsci11081026.

Alsharabi, K. et al. (2022) 'EEG signal processing for Alzheimer's disorders using discrete wavelet transform and machine learning approaches', *IEEE Access*, 10(June), pp. 89781–89797. doi: 10.1109/ACCESS.2022.3198988.

Alsuradi, H., Park, W. and Eid, M. (2020) 'EEG-based neurohaptics research: a literature review', *IEEE Access*, 8, pp. 49313–49328. doi: 10.1109/ACCESS.2020.2979855.

Amezquita-Sanchez, J. P. et al. (2019) 'A novel methodology for automated differential diagnosis of mild cognitive impairment and the Alzheimer's disease using EEG signals', *Journal of Neuroscience Methods*, 322(February), pp. 88–95. doi: 10.1016/j.jneumeth.2019.04.013.

Amini, M. et al. (2021) 'Diagnosis of Alzheimer's disease by time-dependent power spectrum descriptors and convolutional neural network using EEG signal', *Computational and Mathematical Methods in Medicine*, 2021. doi: 10.1155/2021/5511922.

Baglat, P. et al. (2020) 'Multiple machine learning models for detection of Alzheimer's disease using OASIS dataset', *IFIP Advances in Information and Communication Technology*, 617(December), pp. 614–622. doi: 10.1007/978-3-030-64849-7_54.

Batko, K. and Ślęzak, A. (2022) 'The use of big data analytics in healthcare', *Journal of Big Data*. doi: 10.1186/s40537-021-00553-4.

Biagetti, G. et al. (2021) 'Classification of Alzheimer's disease from EEG signal using robust-PCA feature extraction', *Procedia Computer Science*, 192(2019), pp. 3114–3122. doi: 10.1016/j.procs.2021.09.084.

Cassani, R. and Falk, T. H. (2020) 'Alzheimer's disease diagnosis and severity level detection based on electroencephalography modulation spectral "patch" features', *IEEE Journal of Biomedical and Health Informatics*, 24(7), pp. 1982–1993. doi: 10.1109/JBHI.2019.2953475.

Cejnek, M., et al. (November 2021) 'Novelty detection-based approach for Alzheimer's disease and mild cognitive impairment diagnosis from EEG', *Medical & Biological Engineering & Computing*, 59(11–12), pp. 2287–2296. doi: 10.1007/s11517-021-02427-6. Epub 2021 September 18. PMID: 34535856; PMCID: PMC8558189.

Chedid, N. et al. (2022) 'The development of an automated machine learning pipeline for the detection of Alzheimer's disease', *Scientific Reports*, 12(1), pp. 6–12. doi: 10.1038/s41598-022-22979-3.

Cicalese, P. A. et al. (2020) 'An EEG-fNIRS hybridization technique in the four-class classi fi cation of alzheimer' s disease', *Journal of Neuroscience Methods*, 336(January), p. 108618. doi: 10.1016/j.jneumeth.2020.108618.

Counts, S. E. et al. (2017) 'Biomarkers for the early detection and progression of Alzheimer's disease', *Neurotherapeutics*, 14(1), pp. 35–53. doi: 10.1007/s13311-016-0481-z.

De Roeck, E. E. et al. (2019) 'Brief cognitive screening instruments for early detection of Alzheimer's disease: a systematic review', *Alzheimer's Research and Therapy*, 11(1), pp. 1–14. doi: 10.1186/s13195-019-0474-3.

Fakhri, E. et al. (28 January 2023) 'An approach toward artificial intelligence Alzheimer's disease diagnosis using brain signals', *Diagnostics (Basel)*, 13(3), p.477. doi: 10.3390/diagnostics13030477. PMID: 36766582; PMCID: PMC9913919.

Fiscon, G. et al. (2018) 'Combining EEG signal processing with supervised methods for Alzheimer's patients classification', *BMC Medical Informatics and Decision Making*, 18(1), pp. 1–10. doi: 10.1186/s12911-018-0613-y.

Fujita, K. et al. (2022) 'Development of an artificial intelligence- based diagnostic model for Alzheimer's disease', (September), pp. 167–173. doi: 10.1002/agm2.12224.

Goh, C. et al. (2009) 'Comparison of fractal dimension algorithms for the computation of EEG biomarkers for Dementia to cite this version: HAL Id: inria-00442374', *2nd International Conference on Computational Intelligence in Medicine and Healthcare (CIMED2005)*.

He, Y. et al. (2007) 'Regional coherence changes in the early stages of Alzheimer' s disease: a combined structural and resting-state functional MRI study', 35, pp. 488–500. doi: 10.1016/j.neuroimage.2006.11.042.

Hogg, L. and Watt, A. (2012) *Overcoming the stigma of dementia world Alzheimer report 2012*. https://www.alzint.org/u/WorldAlzheimerReport2012ExecutiveSummary.pdf

Houmani, N. et al. (2018) 'Diagnosis of Alzheimer's disease with electroencephalography in a differential framework', *PLoS ONE*, 13(3), pp. 1–19. doi: 10.1371/journal.pone.0193607.

Irankhah, E. (2020) 'Evaluation of early detection methods for Alzheimer's disease', *Bioprocess Engineering*, 4(1), p. 17. doi: 10.11648/j.be.20200401.13.

Kelly, C. J. et al. (2019) 'Key challenges for delivering clinical impact with artificial intelligence', 17(195), pp. 1–9. doi: 10.1186/s12916-019-1426-2.

Knopman, D. S. et al. (2021) 'Alzheimer disease', 7(1), pp. 1–47. doi: 10.1038/s41572-021-00269-y.Alzheimer.

Kumar Shrivastava, A. and Rajak, A. (2020) 'Diagnosis of Alzheimer disease using machine learning approaches', *International Journal of Advanced Science and Technology*, 29(4), pp. 7062–7073. Available at: https://www.researchgate.net/publication/342847161.

Li, L. et al. (2022) 'Potential risk factors of persistent postural - perceptual dizziness: a pilot study', *Journal of Neurology*, 269(6), pp. 3075–3085. doi: 10.1007/s00415-021-10899-7.

Li, W. and Wei, W. (2020) 'Simulation of power spectrum estimation based on AR model', 9(8), pp. 2018–2021. doi: 10.21275/SR20820130325.

Loewenstein, D. A. et al. (2018) 'Novel cognitive paradigms for the detection of memory impairment in preclinical Alzheimer's disease', *Assessment*, 25(3), pp. 348–359. doi: 10.1177/1073191117691608.

Lopes, M., Cassani, R. and Falk, T. H. (8 February 2023) 'Using CNN saliency maps and EEG modulation spectra for improved and more interpretable machine learning-based Alzheimer's disease diagnosis', *Computational Intelligence and Neuroscience*, 3198066. doi: 10.1155/2023/3198066. PMID: 36818579; PMCID: PMC9931465.

McBride, J. C. et al. (2015) 'Sugihara causality analysis of scalp EEG for detection of early Alzheimer's disease', *NeuroImage: Clinical*, 7, pp. 258–265. doi: 10.1016/j.nicl.2014.12.005.

Miltiadous, A. et al. (9 August 2021) 'Alzheimer's disease and frontotemporal dementia: a robust classification method of EEG signals and a comparison of validation methods', *Diagnostics (Basel)*, 11(8), p. 1437. doi: 10.3390/diagnostics11081437. PMID: 34441371; PMCID: PMC8391578

Nobukawa, S. et al. (2020) 'Classification methods based on complexity and synchronization of electroencephalography signals in Alzheimer's disease', *Frontiers in Psychiatry*, 11(April), pp. 1–12. doi: 10.3389/fpsyt.2020.00255.

Oltu, B., Akşahin, M. F. and Kibaroğlu, S. (2021) 'A novel electroencephalography based approach for Alzheimer's disease and mild cognitive impairment detection', *Biomedical Signal Processing and Control*, 63(September 2020). doi: 10.1016/j.bspc.2020.102223.

Pandya, R. et al. (2020) 'Buildout of methodology for meticulous diagnosis of K-complex in EEG for aiding the detection of Alzheimer's by artificial intelligence', *Augmented Human Research*, 5(1). doi: 10.1007/s41133-019-0021-6.

Pegueroles, J. et al. (May 2016) 'Longitudinal brain structural changes in preclinical Alzheimer disease', *Alzheimers Dement*, 13(5), pp. 499–509. doi: 10.1016/j.jalz.2016.08.010. Epub 2016 Sep 28. PMID: 27693189.

Petersen, R. C. et al. (7 September 2010) 'Prevalence of mild cognitive impairment is higher in men The Mayo Clinic Study of Aging', *Neurology*, 75(10), pp. 889–897. doi: 10.1212/WNL.0b013e3181f11d85. PMID: 20820000; PMCID: PMC2938972.

Pirrone, D. et al. (2022) 'EEG signal processing and supervised machine learning to early diagnose Alzheimer's disease', *Applied Sciences (Switzerland)*, 12(11). doi: 10.3390/app12115413.

Puri, D. V et al. (2023) 'Biomedical signal processing and control automatic detection of Alzheimer' s disease from EEG signals using low-complexity orthogonal wavelet filter banks', *Biomedical Signal Processing and Control*, 81(November 2022), p. 104439. doi: 10.1016/j.bspc.2022.104439.

Robert, P. et al. (2020) 'Efficacy of a web app for cognitive training (MEMO) regarding cognitive and behavioral performance in people with neurocognitive disorders: randomized controlled trial', *Journal of Medical Internet Research*, 22(3), pp. 1–11. doi: 10.2196/17167.

Rossini, P. M. et al. (2020) 'Early diagnosis of Alzheimer's disease: the role of biomarkers including advanced EEG signal analysis. Report from the IFCN-sponsored panel of experts', *Clinical Neurophysiology*, 131(6), pp. 1287–1310. doi: 10.1016/j.clinph.2020.03.003.

Safi, M. S., Mohammad, S. and Safi, M. (2021) 'Biomedical signal processing and control early detection of Alzheimer' s disease from EEG signals using Hjorth parameters', *Biomedical Signal Processing and Control*, 65(December 2020), p. 102338. doi: 10.1016/j.bspc.2020.102338.

Subasi, A. (2020) *Use of artificial intelligence in Alzheimer's disease detection, artificial intelligence in precision health*. Elsevier Inc. doi: 10.1016/b978-0-12-817133-2.00011-2.

Subha, D. P. et al. (2010) 'EEG signal analysis: a survey', *Journal of Medical Systems*, 34(2), pp. 195–212. doi: 10.1007/s10916-008-9231-z.

Tavares, G. et al. (2019) 'Improvement in the automatic classification of Alzheimer' s disease using EEG after feature selection', *2019 IEEE International Conference on Systems, Man and Cybernetics (SMC)*, pp. 1264–1269.

Termine, A. et al. (7 April 2021) 'Multi-layer picture of neurodegenerative diseases: lessons from the use of big data through artificial intelligence', *Journal of Personalized Medicine*, 11(4), p. 280. doi: 10.3390/jpm11040280. PMID: 33917161; PMCID: PMC8067806.

Thakare, P. and Pawar, V. R. (2016) 'Alzheimer disease detection and tracking of Alzheimer patient', *Proceedings of the International Conference on Inventive Computation Technologies, ICICT 2016*, 1. doi: 10.1109/INVENTIVE.2016.7823286.

Trambaiolli, L. R. et al. (2017) 'Feature selection before EEG classification supports the diagnosis of Alzheimer's disease', *Clinical Neurophysiology*. doi: 10.1016/j.clinph.2017.06.251.

Trollor, J. N. and Valenzuela, M. J. (2001) 'Brain ageing in the new millennium', *Australian and New Zealand Journal of Psychiatry*, 35(6), pp. 788–805. doi: 10.1046/j.1440-1614.2001.00969.x.

Vecchio, F., Miraglia, F., Alù, F., Menna, M., Judica, E., Cotelli, M., Rossini, P. M. (2020) 'Classification of Alzheimer's disease with respect to physiological aging with innovative EEG biomarkers in a machine learning implementation', *Journal of Alzheimer's Disease*, 75(4), pp. 1253–1261. doi: 10.3233/JAD-200171. PMID: 32417784.

Wang, R. et al. (2015) 'Power spectral density and coherence analysis of Alzheimer's', *Cognitive Neurodynamics*, pp. 291–304. doi: 10.1007/s11571-014-9325-x.

Wang, X. et al. (2023) 'Efficacy assessment of superficial temporal artery—middle cerebral artery bypass surgery in treating moyamoya disease from a hemodynamic perspective: a pilot study using computational modeling and perfusion imaging', *Acta Neurochirurgica*, pp. 613–623. doi: 10.1007/s00701-022-05455-9.

Weston, P. S. J. et al. (2017) 'Serum neurofilament light in familial Alzheimer disease A marker of early neurodegeneration', *Neurology*, 89(21), pp. 2167–2175. doi: 10.1212/WNL.0000000000004667.

Wisniewski, E. D. and Wisniewski, T. (2017) 'Alzheimer's disease: experimental models and reality', *Pathology and Mechanisms of Neurological Disease*, 133(February), pp. 155–175. doi: 10.1007/s00401-016-1662-x.

Woyshville, M. J. and Calabrese, J. R. (1994) 'Quantification of occipital EEG changes in Alzheimer's disease utilizing a new metric: the fractal dimension', *Biological Psychiatryiological Psychiatry*, 35(6), pp. 381–387. doi: doi.org/10.1016/0006-3223(94)90004-3.

Wu, Y. et al. (2017) 'The changing prevalence and incidence of dementia over time—current evidence', (May). doi: 10.1038/nrneurol.2017.63.

You, Z. et al. (2020) 'Alzheimer's disease classification with a cascade neural network', *Frontiers in Public Health*, 8(November), pp. 1–11. doi: 10.3389/fpubh.2020.584387.

Zhang, F. et al. (2019) 'Neurocomputing multi-modal deep learning model for auxiliary diagnosis of Alzheimer' s disease', *Neurocomputing*, 361, pp. 185–195. doi: 10.1016/j.neucom.2019.04.093.

Zhao, X. et al. (2021) 'Application of artificial Intelligence techniques for the detection of Alzheimer's disease using structural MRI images', *Biocybernetics and Biomedical Engineering*, 41(2), pp. 456–473. doi: 10.1016/j.bbe.2021.02.006.

Zhou, J., Yu, J. T., Wang, H. F., Meng, X. F., Tan, C. C., Wang, J., Wang, C. and Tan, L. (2015) 'Association between stroke and Alzheimer's disease: systematic review and meta-analysis', *Journal of Alzheimer's Disease*, 43(2), pp. 479–89. doi: 10.3233/JAD-140666.

9 A Regularized Riemannian Intelligent System for Dementia Screening Using Magnetoencephalography Signals

Srikireddy Dhanunjay Reddy and Tharun Kumar Reddy

9.1 INTRODUCTION

Dementia is a progressive neurological disorder characterized by a decline in cognitive function that interferes with daily activities and is more common in older adults, but it is not a normal part of aging [1–3]. According to the World Health Organization, around 50 million people worldwide have dementia, and there are nearly 10 million new cases every year. Mild cognitive impairment (MCI) is a condition that involves a decline in cognitive abilities that is greater than what is expected for someone's age and education level but not severe enough to interfere significantly with daily activities. People with MCI have a higher risk of developing dementia, but not everyone with MCI will progress to dementia [4, 5]. About 12–16% of those diagnosed with MCI each year develop dementia [6].

Automated early and accurate screening assists persons with appropriate clinical interventions in decreasing the risk factor of MCI to dementia transition. This is both for the purpose of reducing the number of cases in which MCI progresses to dementia as well as for the purpose of reducing the overall number of cases [7, 8]. MC Corsi et al. [9] demonstrated that MEG and electroencephalogram (EEG) biomarkers, when integrated with diagnostic tools, will help in screening dementia stages and other neurological disorders like epilepsy [10], major depressive disorder [11], sleep disorder [12], and driver debility detection [13].

Preprocessing MEG signals consists of a series of processes that aim to clean and ready the data for analysis. The majority of the artifacts that appear in the MEG signal are caused by activities such as fast eye movements, eye blinking, and heart

DOI: 10.1201/9781003479970-9

activity [14]. Data from MEG signal acquisition is of a high dimension and can be properly dealt with in mathematical spaces that are not Euclidean.

As a subfield of differential geometry, Riemannian geometry is concerned with the study of smooth manifolds. It gives a mechanism for defining concepts like length, angle, and curvature on manifolds, which are topological spaces that locally resemble Euclidean space [15, 16]. A Riemannian metric is a mathematical object that gives every smooth curve on a manifold a length and thereby defines the concepts of length and angle in Riemannian geometry. Then, the Riemann curvature tensor is used to characterize the curvature of a Riemannian manifold, which describes the degree to which the manifold is not flat [15, 16]. The highest performance can be attained by feeding the regularized features to the classifiers and tuning the parameters [13, 17–19].

Our analysis of the existing literature on dementia revealed that non-Euclidean approaches can play a major role in improving performance owing to the nonstationary nature of the signal data, but such a paradigm was not much explored in the literature. In addition, MEG signal data has a high spatial resolution in comparison with EEG data and is used in clinical settings for the screening of dementia. Bringing in such high-quality MEG data for research purposes, we present a novel Riemannian-based intelligent system that utilizes MEG signals for highly accurate categorization while demanding few computational resources by reducing redundant information using the pairwise Riemannian distance approach.

In this study, our primary focus was on the categorization of healthy control participants, MCI patients, and dementia patients using regularized Riemannian features. These features exhibit superior performance when compared with other feature extraction methods that were employed in previous classification studies. Further, Section 2 also includes a description of the dataset and problem statement. The proposed solution is then further discussed in Section 3. We compare our detailed dementia classification findings with the previously published research and with the state-of-the-art methods in section 4. We close the chapter with our conclusions including the potential future applications of the suggested approach.

9.2   PROBLEM STATEMENT AND DATASET DESCRIPTION

The primary purpose of this chapter is to perform multiclass classification on the participants, who either were healthy, had MCI, or were diagnosed with dementia. For this study, 160-channel 5 minute resting state MEG signal data was submitted in SPM12 format by Hokuto Hospital, Kumagaya General Hospital, and Mihara Memorial Hospital, all of which are located in Japan. The competition was held as part of the BIOMAG 2022 data analysis conference. Competition organizers were provided both training and testing data, which were captured at two different locations using two different sample frequencies (1000 Hz and 2000 Hz, respectively). For training, we had 100 samples from healthy participants, 29 from persons with dementia, and 15 from people with MCI. Then, for testing, we were given the MEG data of 42 subjects to figure out which class each subject belonged to.

9.3 PROPOSED SOLUTION

In this section, the proposed methodology is broken down into parts and presented in detail, including supporting flow graphs and potential mathematical and analytical explanations. As an overview, the flow of the proposed methodology is shown in Figure 9.1.

The methodology for processing MEG signals involves a sequence of steps aimed at extracting features and classifying the data. The signals with two different frequencies are first down-sampled to a common frequency and then filtered to remove unwanted frequencies. Next, Riemannian geometry feature extraction is used to extract features from the data. These features are then scaled using a standard scaler, which helps to improve the performance of the classifiers. Finally, regularization and classifiers are used to classify the signals based on their features. The objective of this methodology is to identify and analyze patterns in the MEG signals to gain a better understanding of how the brain functions.

9.3.1 Preprocessing

To facilitate analysis, we have included figures depicting raw MEG data from a single participant; Figure 9.2 shows raw MEG data from 160 channels in the time domain, together with details about the data's power spectral density (PSD). The identical subject's PSD and single-channel MEG data are displayed in Figure 9.3.

To preprocess MEG data at two different rates, 1000 Hz and 2000 Hz, we first down-sample the data to 600 Hz to slow down the sample rate and eliminate any high-frequency noise that may be present in the data. The down-sampled data is then sent through a Butterworth filter, which removes frequencies from a signal that are redundant. This step eliminates further noise from the data samples and separates the most important informative frequencies for the study.

After this filtering, the data is often normalized to a common scale to ensure that each epoch is comparable in terms of scale. This is necessary because data with different scales in each epoch can make it difficult to compare results. Next, the data are segmented into smaller 10 second epochs to make the data more manageable and easier to analyze. This segmentation can aid in identifying patterns or trends within the data. The goal of preprocessing is to ready the MEG data for further analysis by eliminating noise and useless frequencies and focusing on the frequencies that are most important for understanding brain activity.

9.3.2 Feature Extraction

In this section, Riemannian geometry-based features are retrieved using the preprocessed signal. These features are then supplied to classifiers, which are responsible for doing the classification. In comparison with Euclidean space, analyzing the electrode location and the correlation between the MEG channels on a Riemannian manifold is a great deal simpler and less complicated. Analyzing the placement of

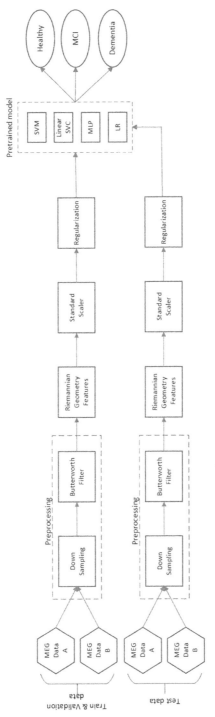

FIGURE 9.1   Workflow of the proposed methodology.

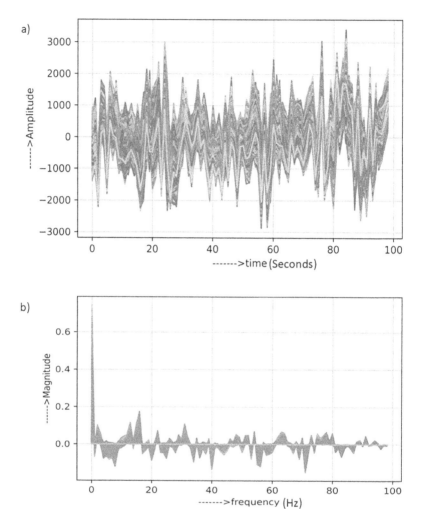

FIGURE 9.2   a) Raw MEG data for 160 channels; b) PSDs of the MEG data.

the MEG electrodes using the Riemannian manifold is recommended for several reasons, one of which is that this method is more effective for analyzing curved surface models. In this chapter, we derived Riemannian covariance matrices through the application of the Ledoit-Wolf covariance estimator, which can be written as

$$A = (1-\alpha)K + \alpha trace(K)I/\eta_a \tag{1}$$

where α in [0,1] is the shrinkage factor, I is the $\eta a X \eta a$ identity matrix, and K is the simple covariance matrix calculated as

$$K = \frac{XX^T}{trace(XX^T)} \tag{2}$$

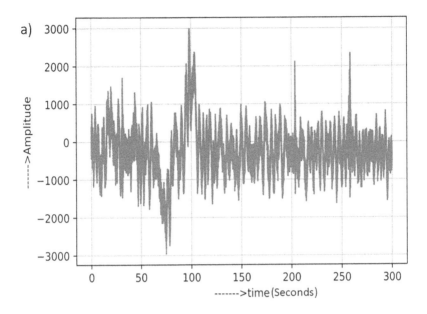

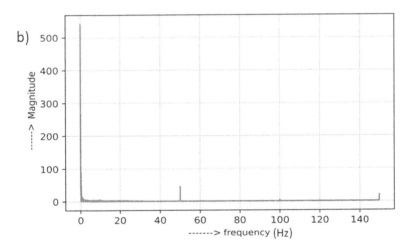

FIGURE 9.3   a) Raw MEG data for a single channel; b) PSD of single-channel MEG data.

The covariance matrix corresponding to i^{th} channel a_i is mapped to a tangent space with the relationship:

$$t_i = upper\left(A_\theta^{\frac{-1}{2}} \, log_\theta \left(A_i \right) A_\theta^{\frac{-1}{2}} \right) \tag{3}$$

Next, we move ahead with the process of channel selection, during which the channels that are found to possess redundant data or data that is not important are eliminated; this is also when the channels are chosen that will be adequate for obtaining the information. Using the Riemannian distance as a criterion, we narrow the list

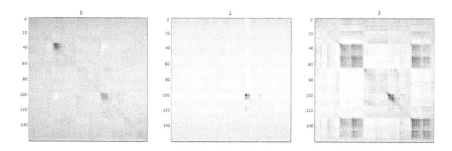

FIGURE 9.4   Covariance matrices of three different classes.

down to 40 channels out of the total of 160 data channels at this point to measure the mean threshold Riemannian distance:

$$A_{tres} = \sum_{i=1}^{N} \sum_{j>n}^{N} \delta_R \left(\overline{A}^{(i)}, \overline{A}^{(j)} \right) \tag{4}$$

where K = 2 denotes the number of categories and δ_R denotes pairwise Riemannian distance, calculated as

$$\delta_R = \left\| Log \left(A_1^{-1} A_2 \right) \right\|_F = \sqrt{\sum_{n=1}^{N_e} log^2 \lambda_n} \tag{5}$$

where Log(.) is the log-matrix operator and ll.llF is the Frobenius norm of a matrix. Following the selection of the channels, we will need to provide the classifier with these attributes in the form of input parameters. These features are then fed into tangent space blocks so that they can be converted from Riemannian space to Euclidean space in order to accomplish this goal. Figure 9.4 represents the Riemannian covariance matrices of three different class subjects using the discussed technique.

9.3.3   REGULARIZATION

In this section, regularization is performed following standard scaling, which is performed after the Riemannian covariance matrix features are converted from tangent space to Euclidean space. Regularization refers to a collection of approaches used in machine learning and statistical modeling to prevent overfitting and improve model generalization. Overfitting happens when a model is overly complicated and captures too much noise in the data, resulting in poor performance on new data. There are different types of regularization techniques.

L1 regularization (Lasso): L1 regularization adds a penalty term proportional to the absolute value of the model coefficients to the cost function. This leads to sparse models where some of the coefficients are set to zero. The L1 regularization cost function can be formulated as follows,

$$J(\theta) = MSE(\theta) + \alpha \sum |\theta_j| \tag{6}$$

where $J(\theta)$ is the regularized cost function, $MSE(\theta)$ is the mean squared error cost function, α denotes the regularization parameter that controls the strength of penalty term, and θ_j indicates the jth model coefficient.

L2 regularization (Ridge): L2 regularization adds a penalty term proportional to the square of the model coefficients to the cost function. This leads to smoother models with smaller coefficients. The L2 regularization cost function can be formulated as follows:

$$J(\theta) = MSE(\theta) + \alpha \sum (\theta_j)^2 \tag{7}$$

Elastic net: Elastic net regularization is a technique that combines both L1 (Lasso) and L2 (Ridge) regularization. It was introduced to balance the strengths and weaknesses of both L1 and L2 regularization. Similar to L1 and L2 regularization, elastic net regularization also adds a penalty term to the cost function. The penalty term is a combination of the L1 and L2 penalty terms. The elastic net regularization cost function can be formulated as follows:

$$J(\theta) = MSE(\theta) + \alpha 1 \sum |\theta_j| + \alpha 2 \sum (\theta_j)^2 \tag{8}$$

Elastic net regularization combines the sparsity-inducing property of L1 regularization with the smoothness-inducing property of L2 regularization. It is particularly useful when there are many correlated features in the data. L1 regularization tends to select one of the correlated features and ignore the others, while L2 regularization tends to keep all the correlated features. Elastic net regularization can balance between these two extremes and select a subset of the correlated features. The values of $\alpha 1$ and $\alpha 2$ can be chosen by cross-validation to find the optimal trade-off between sparsity and smoothness. In general, elastic net regularization is a powerful regularization technique that can handle high-dimensional datasets and prevent overfitting.

The regularization approach used is determined by the specific challenge and the features of the data. Regularization can help improve model accuracy and lessen the danger of overfitting. It can also be used to mitigate the effects of multicollinearity in linear regression models. Regularization can also have certain disadvantages, such as increasing the computational complexity of the model and making coefficient interpretation more difficult. In summary, regularization is a useful strategy in machine learning and statistical modeling for preventing overfitting and improving model generalization. Different regularization procedures have advantages and disadvantages and should be selected based on the individual data characteristics. When comparing the performances of L1, L2, and elastic net regularization on our unique nonstationary MEG dataset, the L2 strategy produced good results.

9.3.4 CLASSIFICATION

The main contribution of this section is to know the best-performing classifier for the given MEG signal data and make the model to train for the accurate classification of test data samples. MEG signal classification is an important task in neuroscience that involves using machine learning algorithms to automatically categorize MEG signals into predefined categories or classes. Classifiers such as decision trees, k-nearest neighbor (KNN), support vector machine (SVM), and artificial neural networks can be used to identify patterns and relationships within MEG data, allowing us to better understand brain activity and make predictions about cognitive processes.

The success of MEG signal classification depends on several factors, such as the quality and quantity of the data used to train the classifier, the choice of features or attributes used to describe the MEG signals, and the algorithm and its parameters. Evaluating the performance of a classifier is crucial in ensuring its accuracy, and metrics such as sensitivity, specificity, and F1 score are commonly used to assess how well the classifier can differentiate between different types of MEG signals. MEG signal classification has many applications in neuroscience research, such as identifying different stages of sleep or detecting neural responses to specific stimuli. However, the complexity of MEG data and the challenges of data preprocessing and feature extraction make MEG signal classification a challenging task. Nevertheless, with careful design, training, and evaluation of classifiers, MEG signal classification has the potential to greatly enhance our understanding of the brain and its functioning.

9.4 RESULTS

9.4.1 COMPARISON AMONG CLASSIFIERS

A holistic comparison of classifiers indicates that models that are built upon Riemannian geometry-based features provide the most accurate predictions of dementia screening using MEG signals. The collected features were fed into all of the available baseline classifiers, and the parameters of those classifiers were fine-tuned to achieve the best possible results. The results of training and testing sets for SVM with and without the mutual information feature selection (MIFS) algorithm, as well as multilayer perceptron (MLP), are shown in Figure 9.5. Some classifiers' tuned parameters were examined in detail and are discussed here.

> **KNN:** This model's hyperparameters include class weights and nearest neighbors, while feature inputs consist of tangent space vectors.
> **Linear SVC:** This model was supplied with covariance and tangent space vectors as input vectors with an L2 penalty tolerance of 0.05 and a regularization parameter of 0.6.
> **Linear Regression:** This model was fed tangent space vectors as inputs, and the optimal performance was attained with an elastic net penalty and regularization terms between 0.1 and 0.6.
> **MLP:** The model's inputs were covariance matrices, and its parameters were fine-tuned. With two concealed layers of 'Relu' adaptive learning rate

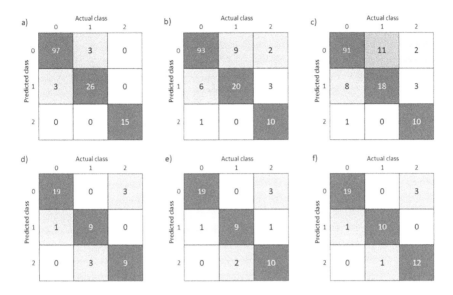

FIGURE 9.5 Confusion matrices of training & testing data: a) & d) SVM; b) & e) SVM+MIFS; c) & f) MLP.

activation and the maximum number of iterations until model optimization, the model achieved its highest performance.

SVM: SVM is optimized by balancing class weights ($0.1 \leq C \leq 1$). The regularization parameters are automatically tuned, and the tangent space vectors serve as feature inputs to the classifier.

Fivefold cross-validation was used to quantify and compare the effectiveness of the classifiers, and the results are depicted graphically in Figure 9.6. This method can be used to focus on the optimal classifier for a specific dataset and application.

9.4.2 COMPARISON WITH OTHER STATE-OF-THE-ART WORKS

In this section, we evaluate the effectiveness of the proposed methodology in comparison to other recent publications that have been published. Table 9.1 makes it abundantly evident that our method of categorizing dementia is one of a unique kind and highly effective owing to the utilization of Riemannian features in conjunction with the fine-tuning of classifier parameters. With an accuracy in multiclass classification that was 92.11%, the new method beat all of the previous ones at dementia screening. A graphical comparison of the works that were tabulated could be seen in Figure 9.7.

9.5   CONCLUSIONS AND FUTURE WORK

With the help of this research, we came up with a novel method of Riemannian geometry covariance matrices for the categorization of dementia based on MEG

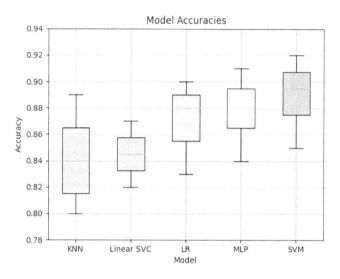

FIGURE 9.6 Comparison between different classifiers for fivefold cross-validation.

TABLE 9.1
Performance Comparison of Dementia Screening

Authors	Feature extraction	Modality	Classes	Number of subjects	Accuracy (%)
Bruna et al. [20]	Statistical	MEG	MCI vs Healthy	MCI: 26 Healthy: 18	65
Ahmadlou et al. [21]	Wavelet transform	EEG	Alzheimer vs Healthy	Alzheimer: 20 Healthy: 7	90.3
Mannone et al. [22]	Fourier transform	EEG	Alzheimer vs MCI	Alzheimer: 7 MCI: 25	*
Houmani et al. [23]	Entropy	MEG	Alzheimer vs Healthy	Alzheimer: 30 Healthy: 19	83
Amezquita-Sanchez et al. [24]	Entropy	EEG	Alzheimer vs MCI vs Healthy	Healthy: 45 MCI: 45 Alzheimer: 45	86.9
Hornero et al. [25]	Spectral analysis	MEG	Alzheimer vs Healthy	Alzheimer: 20 Healthy: 21	80.5
Morabito et al. [26]	Continuous wavelet transform	MEG & EEG	Alzheimer vs MCI vs Healthy	Alzheimer 60 MCI: 56 Healthy: 23	78
Leracitano et al. [27]	Power spectral density	EEG	Alzheimer vs MCI vs Healthy	Alzheimer: 63 MCI: 63 Healthy: 63	83.3
This work	Riemannian features	MEG	Dementia vs Healthy vs MCI	Healthy: 125 Dementia: 40 MCI: 22	92.11

Note: * - Not available in the published manuscript.

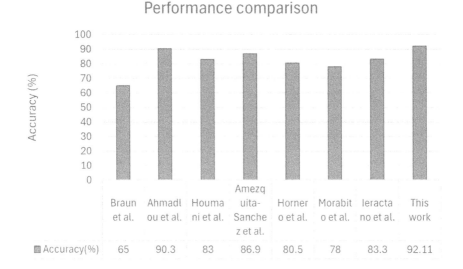

FIGURE 9.7   Comparison of works on dementia screening.

signals, which achieved the highest accuracy of 92.11% utilizing 42 individuals as test subjects. When applying the idea of pairwise Riemannian distance, redundant information channels were removed, which reduced computational cost and complexity. Along with tangent space vectors, covariance matrices of Riemannian geometry were required to improve the performance of some classifiers; however, this circumstance increases the amount of computational work that must be done. To address this specific issue, future work includes implementing adaptive Riemannian methods to reduce the functional variability of models, exploring the use of Riemannian geometry for electrode selection through spatio-spectral filters, and extending the proposed methods to other modalities such as EEG and fMRI, as well as applying the concept to deep learning models.

REFERENCES

1. P. K. Panegyres, R. Berry, J. Burchell, Early dementia screening, Diagnostics 6 (1) (2016) 6.
2. J.-H. Chen, K.-P. Lin, Y.-C. Chen, Risk factors for dementia, Journal of the Formosan Medical Association 108 (10) (2009) 754–764.
3. M. F. Mendez, J. L. Cummings, *Dementia: a clinical approach*, Butterworth-Heinemann, 2003.
4. N. Serrano, D. López-Sanz, R. Bruña, P. Garcés, I. Rodríguez-Rojo, A. Marcos, D. P. Crespo, F. Maestú, Spatiotemporal oscillatory patterns during working memory maintenance in mild cognitive impairment and subjective cognitive decline, International Journal of Neural Systems 30 (1) (2020) 1950019.
5. S. Gauthier, B. Reisberg, M. Zaudig, R. C. Petersen, K. Ritchie, K. Broich, S. Belleville, H. Brodaty, D. Bennett, H. Chertkow, et al., Mild cognitive impairment, The lancet 367 (9518) (2006) 1262–1270.

6. A. J. Mitchell, M. Shiri-Feshki, Rate of progression of mild cognitive impairment to dementia–meta-analysis of 41 robust inception cohort studies, Acta psychiatrica scandinavica 119 (4) (2009) 252–265.

7. F. Miraglia, F. Vecchio, C. Marra, D. Quaranta, F. Al'u, B. Peroni, G. Granata, E. Judica, M. Cotelli, P. M. Rossini, Small world index in default mode network predicts progression from mild cognitive impairment to dementia, International Journal of Neural Systems 30 (2) (2020) 2050004.

8. W. Feng, N. V. Halm-Lutterodt, H. Tang, A. Mecum, M. K. Mesregah, Y. Ma, H. Li, F. Zhang, Z. Wu, E. Yao, et al., Automated MRI-based deep learning model for detection of Alzheimer's disease process, International Journal of Neural Systems 30 (6) (2020) 2050032.

9. M.-C. Corsi, M. Chavez, D. Schwartz, L. Hugueville, A. N. Khambhati, D. S. Bassett, F. De Vico Fallani, Integrating EEG and MEG signals to improv motor imagery classification in brain–computer interface, International Journal of Neural Systems 29 (1) (2019) 1850014.

10. P. P. M. Shanir, K. A. Khan, Y. U. Khan, O. Farooq, H. Adeli, Automatic seizure detection based on morphological features using one-dimensional local binary pattern on long-term EEG, Clinical EEG and Neuroscience 49 (5) (2018) 351–362.

11. W. Mumtaz, S. S. A. Ali, M. A. M. Yasin, A. S. Malik, A machine learning framework involving EEG-based functional connectivity to diagnose major depressive disorder (MDD), Medical & Biological Engineering & Computing 56 (2018) 233–246.

12. C. K. Behera, T. K. Reddy, L. Behera, B. Bhattacarya, Artificial neural network based arousal detection from sleep electroencephalogram data, in: 2014 International Conference on Computer, Communications, and Control Technology (I4CT), IEEE, 2014, September, pp. 458–462.

13. T. K. Reddy, V. Arora, V. Gupta, R. Biswas, L. Behera, EEG-based drowsiness detection with fuzzy independent phase-locking value representations using lagrangian-based deep neural networks, IEEE Transactions on Systems, Man, and Cybernetics: Systems 52 (1) (2021) 101–111.

14. A. H. Treacher, P. Garg, E. Davenport, R. Godwin, A. Proskovec, L. G. Bezerra,. . .& A. A. Montillo, Megnet: automatic ICA-based artifact removal for meg using spatiotemporal convolutional neural networks, NeuroImage 241 (2021) 118402.

15. S. Gallot, D. Hulin, J. Lafontaine, *Riemannian geometry*, Vol. 2, Springer, 1990.

16. M. P. Do Carmo, J. Flaherty Francis, *Riemannian geometry*, Vol. 6, Springer, 1992.

17. H. Jin, K. Ranasinghe, P. Prahu, C. Dale, Y. Gao, K. Kudo, K. Vossel, A. Raj, S. Nagarajan, F. Jiang, Dynamic functional connectivity meg features of alzheimer's disease, bioRxiv (2023) 2023–02.

18. A. Tiwari, A. Chaturvedi, Automatic EEG channel selection for multiclass brain–computer interface classification using multi objective improved firefly algorithm, Multimedia Tools and Applications 82 (4) (2023) 5405–5433.

19. T. K. Reddy, Y.-K. Wang, C.-T. Lin, J. Andreu-Perez, Joint approximate diagonalization divergence based scheme for EEG drowsiness detection brain computer interfaces, in: 2021 IEEE International Conference on Fuzzy Systems (FUZZ-IEEE), IEEE, 2021, pp. 1–6.

20. R. Bruna, J. Poza, C. Gómez, M. Garcia, A. Fernández, R. Hornero, Analysis of spontaneous MEG activity in mild cognitive impairment and Alzheimer's disease using spectral entropies and statistical complexity measures, Journal of Neural Engineering 9 (3) (2012) 036007.

21. M. Ahmadlou, H. Adeli, A. Adeli, Fractality and a wavelet-chaosmethodology for EEG-based diagnosis of Alzheimer disease, Alzheimer Disease & Associated Disorders 25 (1) (2011) 85–92.

22. N. Mammone, L. Bonanno, S. D. Salvo, S. Marino, P. Bramanti, A. Bramanti, F. C. Morabito, Permutation disalignment index as an indirect, EEG based, measure of brain connectivity in MCI and AD patients, International Journal of Neural Systems 27 (5) (2017) 1750020.

23. N. Houmani, F. Vialatte, E. Gallego-Jutgl'a, G. Dreyfus, V.-H. Nguyen- Michel, J. Mariani, K. Kinugawa, Diagnosis of Alzheimer's disease with electroencephalography in a differential framework, PLoS ONE 13 (3) (2018) e0193607.

24. J. P. Amezquita-Sanchez, N. Mammone, F. C. Morabito, H. Adeli, A new dispersion entropy and fuzzy logic system methodology for automated classification of dementia stages using electroencephalograms, Clinical Neurology and Neurosurgery 201 (2021) 106446.

25. R. Hornero, J. Escudero, A. Fernández, J. Poza, C. Gómez, Spectral and nonlinear analyses of meg background activity in patients with Alzheimer's disease, IEEE Transactions on Biomedical Engineering 55 (6) (2008) 1658–1665.

26. F. C. Morabito, M. Campolo, C. Ieracitano, J. M. Ebadi, L. Bonanno, A. Bramanti, S. Desalvo, N. Mammone, P. Bramanti, Deep convolutional neural networks for classification of mild cognitive impaired and Alzheimer's disease patients from scalp EEG recordings, in: 2016 IEEE 2nd International Forum on Research and Technologies for Society and Industry Leveraging a Better Tomorrow (RTSI), IEEE, 2016, pp. 1–6.

27. C. Ieracitano, N. Mammone, A. Bramanti, A. Hussain, F. C. Morabito, A convolutional neural network approach for classification of dementia stages based on 2D-spectral representation of EEG recordings, Neurocomputing 323 (2019) 96–107.

10 Detecting Dementia Using EEG Signal Processing and Machine Learning

Mahbuba Ferdowsi, Choon-Hian Goh,
Gary Tse, and Haipeng Liu

10.1 INTRODUCTION

Dementia is a syndrome associated with an ongoing decline in brain functioning, with impaired ability to remember, think, or make decisions. It is an increasing global public health problem, with nearly 10 million new cases each year. The total number of people with dementia is projected to reach 82 million in 2030 and 152 million in 2050 (1). Dementia imposes a huge economic burden, with the costs are associated with the severity (2). The early detection of dementia plays a key role in improving the intervention and management and reducing the costs.

Dementia is an umbrella term for various illnesses, all of which involve neurodegenerative changes in the brain. Some of the commonest types of dementia include Alzheimer's disease, vascular dementia, frontotemporal dementia, and Lewy body dementia, while there are more than 100 rarer causes (Figure 10.1). Different types of dementia differ in pathology, features, symptoms, and treatment (3). Finally, in some cases mixed dementia is observed. Growing evidence suggests a continuum between neurodegeneration and vascular dysfunction. The distinction between isolated Alzheimer's disease, vascular dementia, and mixed dementia remains controversial and challenging in clinical practice (4).

Many risk factors of dementia have been identified, e.g., uncontrolled hypertension (5), diabetes (6), hypercholesterolemia (7), excessive alcohol, and traumatic brain injury, which is particularly important in early-onset dementia, as well as some genetic factors (8). Some risk factors also play important roles in cardiovascular diseases in the aging process (9) (Figure 10.2). Various predictive factors change their roles along with ages and stages of dementia, where many factors are modifiable (10). Evidence shows that modifying 12 key risk factors might prevent or delay up to 40% of dementias (8).

Despite accumulative observation of the epidemiology, symptoms, and risk factors of dementia (11, 12), there is a lack of in-depth understanding of its pathology. Though lab tests and neuroimaging can afford reliable diagnosis of dementia, they

DOI: 10.1201/9781003479970-10

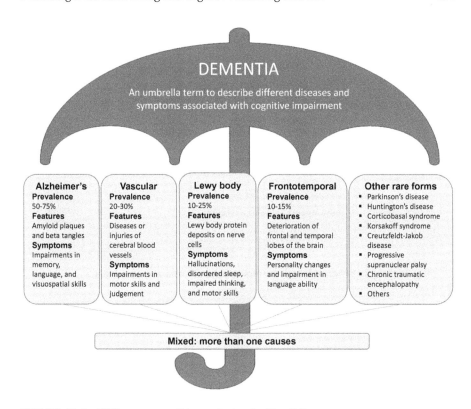

FIGURE 10.1   Different types of dementia, inspired by (16).

are costly and dependent on professional operations and therefore difficult to apply as screening tools (13). In clinical practice, patient performance-based testing and informant interviews are commonly used tools for dementia screening, but they are not reliable in some cases (14). For example, many cognitive screening tests are available for Alzheimer's disease, but most tests are only validated in a memory clinic setting and not standardized (15). As a result, the accurate early detection of dementia is still challenging.

10.2   ELECTROENCEPHALOGRAM: A POTENTIAL TOOL FOR DEMENTIA DETECTION

Electroencephalography (EEG) is a technology that records the spontaneous electrical activity of the brain using electrodes attached on the scalp. Dementia-associated changes in brain electrophysiology can lead to changes in various EEG features in the time, frequency, and time–frequency domains, providing new bio-markers for the diagnosis of dementia (17). EEG signals have been applied in the diagnosis of various brain disorders such as epilepsy, tremor, concussions, strokes, and sleep disorders.

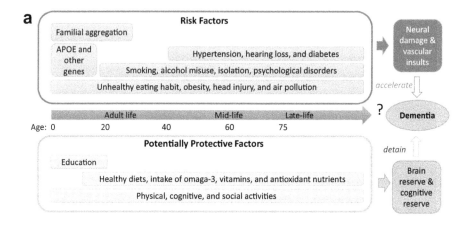

Risk factors for atherosclerotic cardiovascular disease and dementia

b

Smoking*

Triglycerides*

The gut microbiome*

Body mass index*

LDL and total cholesterol*

NAFLD*

Dementia

HDL cholesterol*

Hypertension

Physical inactivity

Diet

Genetics*

Diabetes

FIGURE 10.2 Risk factors for dementia. a: Risk factors in different stages, inspired by (10). b: Shared risk factors between dementia and atherosclerotic cardiovascular disease, adapted from (**9**). APOE: Apolipoprotein E, LDL: low-density lipoprotein, HDL: high-density lipoprotein, NAFLD: non-alcoholic fatty liver disease.

However, EEG recordings often contain different frequency components with large data size. In addition, EEG often has low spatial resolution and poor signal-to-noise ratio. To improve the efficiency in EEG signal analysis, artificial intelligence (AI) methods, especially machine learning, have been developed for the classification (18). Machine learning can effectively reduce data dimensions and integrate multimodal clinical data to improve the accuracy and reliability of EEG-base diagnosis. Recently, machine learning-assisted EEG analysis has been applied in the early detection of dementia. We aim with this chapter to provide a comprehensive review of the state of the art to summarize the advantages, limitations, and some future directions.

10.3 APPLYING AI IN EEG-BASED DEMENTIA DETECTION

The identification and evaluation of dementia can be aided by AI (19). Researchers have utilized AI to develop clinical decision-making systems, find novel treatments, and study the genomes of dementia patients (20). They have accomplished this utilizing various biomarkers that can be measured to indicate a biological state of cognitive decline or dementia, such as neuroimaging, retinal imaging, language information, cerebrospinal and blood biomarkers, and genetic information (21). AI has achieved significant advancements in processing images and natural language, leading to the development of speech-to-speech translation engines and spoken dialogue systems (22). These advancements enable a faster and more sophisticated analysis of neuroimaging and speech data (23, 24). In the case of complex data sources like magnetic resonance imaging, positron emission tomography (PET) neuroimaging, and cerebrospinal fluid biomarkers, deep learning methods such as neural networks can be utilized to develop diagnostic classifiers or feature extraction techniques. AI has the capability to revolutionize the early identification and monitoring of dementia, allowing medical practitioners to step in early and improve patient outcomes (25, 26). Table 10.1 provides a summary of recent studies that have applied AI to diagnose dementia, and next we give details of the algorithms used in these studies.

10.3.1 Extracting Multidomain EEG Features

10.3.1.1 Time

To perform time-domain analysis, some of the methods based on wavelet theory could be utilized on dementia research included continuous wavelet transform (CWT), wavelet packet decomposition (WPD), and wavelet packet transform (WPT) (27–29). Using CWT, a signal can be gradually broken down into its individual frequencies. CWT employs a variety of wavelet functions that can be scaled and shifted to correspond to the frequency and time scale of the signal under study. This enables CWT to analyze nonstationary signals, in which the statistical characteristics of the signal may change over time. It has been used in dementia research to examine EEG data and spot variations in brain activity over time in people with Alzheimer's disease or mild cognitive impairment.

WPD is a method that breaks down a signal into its component wavelet packets, which are sub-bands of the original signal at various frequency ranges. Because it splits the signal into more wavelet packets than CWT, WPD enables a more thorough examination

TABLE 10.1

Applications of AI in Dementia Detection

Algorithm	References	No. of subjects and their class	Feature extraction techniques	Features	Performance	Strengths	Limitations
SVM	(Wang et al., 2019) (31)	(VaD = 15, HC = 21); VaD vs HC	Directed transfer function	Interregional connections	Acc: 0.86 Sen: 0.87 Spe: 0.86 AUC: 0.85	Resistant to noise and outliers and is capable of handling such data.	1. Limited number of subjects, 36, 47, and 44, respectively. 2. Operate most effectively when the number of features is modest and when the features are carefully chosen.
	(Sharma et al., 2019) (28)	(MCI = 16, Dementia = 15, HC = 13) 1. HC vs MCI 2. HC vs Dementia 3. MCI vs Dementia	Wavelet packet transform and power spectral density	PSD spectral features, spectral entropy, skewness, kurtosis, spectral skewness, spectral crest factor, fractal dimension.	In FTT state: 1. Acc: 0.90 2. Acc: 0.84 3. Acc: 0.84 In CPT state: 1. Acc: 0.74 2. Acc: 0.85 3. Acc: 0.88		
	(Sharma, Kolekar and Jha, 2021) (29)	(MCI = 16, Dementia= 16, HC = 15); Dementia vs MCI vs HC	Wavelet packet decomposition and power spectral density	Skewness, kurtosis, spectral kurtosis, spectral entropy, fractal dimension and PSD spectral features	MST: Acc: 0.92, 0.92, and 0.95 respectively.		
	(Trambaiolli et al., 2011) (57)	(AD = 16, HC = 19); AD vs HC	Spectral Analysis	Coherence	Acc: 0.80 Sen: 0.832		

	Study	Dataset	Method	Features	Results	Notes
KNN	(Durongbhan et al., 2019) (27)	(AD = 20, HC = 20); AD vs HC	Continuous Wavelet Transform (CWT), frequency domain, and time-frequency domain.	FFT coefficient and CWT coefficient.	Acc: 0.99	1. Spectral analysis reveals the frequency structure of the EEG signal, but they could not shed light on underlying physiological processes that give rise to the reported spectral patterns. 2. Less relevant in high-dimensional spaces, and less precise as well.
	(Jennings et al., 2022) (33)	(AD = 32, PDD = 22, DLB = 26, HC = 18); HC vs PDD AD vs DLB	Spectral analysis and Welch's Periodogram	Dominant frequency and dominant frequency variance.	1. Spe: 0.87 Sen: 0.92 2. Spe: 0.75 Sen: 0.91	1. Can do binary and multi-class classification problems. 2. non-parametric method that makes no assumptions regarding the spatial distribution of the data.
	(Sharma et al., 2020) (38)	(Dementia = 16, Early Dementia = 16, HC = 15); Dementia vs early Dementia vs HC	Iterative filtering decomposition	PSD, variance, Fractal dimension, entropy	Acc: 0.92 vs 0.916 vs 0.918	

(Continued)

TABLE 10.1 (*Continued*)
Applications of AI in Dementia Detection

Algorithm	References	No. of subjects and their class	Feature extraction techniques	Features	Performance	Strengths	Limitations
RF	(Dauwan et al., 2016) (40)	(DLB = 66, AD = 66, HC = 66; 1. DLB vs. AD. DLB vs. HC. 3. AD vs. HC.	(N/A)	Non-EEG features (lowest delta power) and clinical features (hallucinations)	1. Acc: 0.87 Sen: 0.88 Spe: 0.86 2. Acc: 0.94 Sen: 0.95 Spe:0.93 3. Acc: 0.91 Sen: 0.92 Spe: 0.91	Are less prone to overfitting than other machine learning models like SVM.	Can be highly computational and take prolonged processing times.
	(Miltiadous et al., 2021) (52)	(AD = 10, FTD = 10, HC = 8); 1. AD vs HC. 2. FTD vs HC. 3. AD vs FTD	Spectral analysis	Mean, variance, and IQR	1. Acc:0.991 Sen: 0.986 Spe:0.99 2. Acc: 0.98 Sen: 0.98 Spe: 0.98 3. Acc: 0.977 Sen: 0.978 Spe: 0.975		
Neural network	(So et al., 2017) (47)	(Dementia = 573, MCI = 663, Cognitive decline = 2965, HC=9799); 1. HC vs cognitive decline 2. MCI vs Dementia	(N/A)	Orientation to place, orientation to time, three-stage commands, recall, attention, repetition, registration, complex commands, language, age	1. TP rate: 0.97 F measure: 0.97 2. TP rate: 0.74 F measure: 0.74	Can shed light on the characteristics for classification, which can help with the interpretation and comprehension of underlying mechanisms.	Fails to perform well in tasks that require spatial data.

	Reference	Method	Dataset	Features	Results	Notes
	(Ieracitano et al., 2020) (30)	Time- frequency analysis and bispectrum (BiS) analysis.	(AD = 63, MCI = 63, HC = 63); 1. AD vs HC. 2. AD vs MCI. 3. MCI vs HC. 4. AD vs MCI vs HC	CWT features vector, BiS features vector and multimodal (CWT + BiS) features vector.	1. Acc: 0.96 Precision: 0.93 Recall: 0.93 2. Acc: 0.87 Precision: 0.83 Recall: 0.81 3. Acc: 0.92 Precision: 0.91 Recall: 0.90 4. Acc: 0.82 Precision: 0.73 Recall: 0.68	
	(Anuradha & Jamal, 2021) (32)	FFT	(DLB = 12 HC = 18); DLB vs HC	Dominant frequency, mean dominant frequency, dominant frequency variability, frequency prevalence.	Acc: 0.944, Sen:0.909, Spe:1.	
Deep learning	(Ieracitano et al., 2019) (36)	(N/A)	(AD = 63, MCI = 63, HC = 63) 1. AD vs. HC. 2. AD vs MCI 3. MCI vs HC 4. AD vs. MCI vs HC	PSD spectral representation.	1. Acc: 0.93 Precision: 0.91 Recall: 0.95 2. Acc: 0.85 Precision: 0.84 Recall: 0.85 3. Acc: 0.92 Precision:0.92 Recall: 0.92 4. Acc: 0.83 Precision: 0.80 Recall: 0.83	Customized models produced greater accuracy and more performance optimized. Processing power is high, especially if they include deep structures with multiple layers.

(Continued)

TABLE 10.1 (Continued)
Applications of AI in Dementia Detection

Algorithm	References	No. of subjects and their class	Feature extraction techniques	Features	Performance	Strengths	Limitations
	(Ieracitano et al., 2020) (58)	(AD = 63, MCI = 63, HC = 63); 1. AD vs. HC 2. AD vs MCI 3. MCI vs HC	(N/A)	(N/A)	1. Acc: 0.858 ± 0.22 2. Acc: 0.690 ± 0.13 3. 0.853 ± 0.18		
	(Xia et al., 2023) (45)	(AD = 49, MCI = 37, HC = 14 AD vs MCI vs HC	FFT	(N/A)	Acc: 0.971 F1 score: 0.971		
	(Amini et al., 2021) (59)	(AD = 64, MCI = 64, HC = 64) MCI vs AD vs HC	Time-dependent power spectrum descriptors	(N/A)	Acc: 0.82.3 AUC: 0.988		

Note: N/A, not applicable; HC, healthy control; VaD, Vascular dementia; MCI, mild cognitive impairment; AD, Alzheimer's disease; FTD, frontotemporal dementia; DLB, dementia with Lewy bodies; PDD, Parkinson's disease dementia; NL, normal cognition; CU, cognitively unpaired; MST, motor speed test; PSD, power spectral density; FTT, finger tapping test; CPT, continuous performance test; Acc, accuracy; Sen, sensitivity; AUC, area under curve; FFT, fast Fourier transform; SVM, support vector machine; KNN, K-nearest neighbors; RF, random forest; NN, neural network; CNN, convolutional neural network.

of the signal. WPD has been utilized in dementia research to pinpoint particular frequency bands linked to alterations in cognitive function and behavior throughout time. WPT is comparable with WPD, except it breaks down the signal into a collection of orthogonal basis functions as opposed to wavelet packets. WPT is an additional computationally efficient variations of WPD that can be used to analyze signals with many data points. WPT has been applied in dementia research to find subgroups of people who have shared cognitive trajectories across time. Given that bispectrum (BiS) estimation needs many observations of the signal, BiS analysis is frequently employed to analyze signals in the time domain (30). It is especially helpful for researching nonlinear signal interactions, such as those that might take place in the brain during cognitive processing.

10.3.1.2 Frequency

The following methods can be used for frequency domain analysis: directed transfer function, fast Fourier transform (FFT) and Welch's periodogram (31–33). When examining the directed information flow between various frequency components in a signal, discrete-time Fourier (DTF) frequency analysis is used. The multivariate autoregressive model, which calculates the causal links between the various elements of a multivariate time series, serves as the foundation for the DTF.

FFT is a popular frequency domain analysis method for breaking down a time-domain signal into its individual frequency components that offers details on each frequency component's amplitude and phase. FFT is frequently employed to detect the existence of frequency elements in a signal, including those linked to various cognitive functions in dementia detection. The power spectral density (PSD) of a signal can be calculated using Welch's periodogram. By separating the signal into overlapping segments, this method minimizes the variance of the PSD estimate, making it an improvement over the traditional periodogram method. For signals with nonstationary characteristics, Welch's periodogram is especially helpful. Compared with other approaches, it offers a more precise PSD estimate.

10.3.1.3 Time–Frequency Analysis

Wavelet analysis is a commonly used technique to obtain the EEG features in the time–frequency domain. The Montreal Neurological Institute (MNI) model can be used to normalize and co-register EEG data from several people onto a single brain region in time–frequency analysis (34). This enables group-level study of time–frequency properties like oscillatory power, coherence, and event-related spectral disturbances. Additionally, the MNI model allows for source localization of time–frequency activity, which can be used to pinpoint the precise brain areas that are involved in various cognitive processes and impairment.

10.3.1.4 Other Advanced Features

In addition to the traditional EEG features, advanced features can be extracted from the information theory domain such as Shannon entropy (35). These measures can provide insights into the complexity and organization of brain activity in dementia patients. As EEG electrodes are distributed across the scalp, two-dimensional features can be extracted to explore the interconnections among different brain regions during the development of dementia (36).

Analyzing the spatial patterns of EEG activity can help identify abnormalities or disruptions in the functional connectivity between brain regions. Nonlinear dynamic techniques have also been utilized to analyze EEG signals in dementia. Since brain neurons exhibit nonlinear behavior, techniques that capture nonlinear phenomena, such as threshold and saturation, can provide valuable information. Changes in the nonlinearity of neural activities detected through EEG can indicate the presence of dementia (37). Furthermore, studies have reported fewer fractal dimensions in the parietal and temporal cortices of patients with Alzheimer's disease compared with those of healthy subjects (28, 38). Fractal-based metrics computed from resting-state EEG have been proposed as potential indicators of different stages of dementia progression (39).

10.3.2 EXTRACTED FEATURES IN RECENT WORKS

The features in Table 10.1 include statistical, spectral, non-EEG, and clinical ones. Skewness and kurtosis are statistical features. A distribution's skewness indicates its asymmetry, where positive and negative skewness denote a larger tail on the right and left sides, respectively; skewness can be used to examine the form of neuroimaging measurements like grey matter volume or thickness of cortical in dementia. Kurtosis reflects the flatness of a distribution, where high kurtosis indicates a strong peak and long tails.

PSD spectral features, spectral entropy, spectral skewness, and spectral crest are among the features derived from spectral analysis (29). PSD spectral features are statistical evaluations of a signal's power distribution throughout several frequency bands. The complexity or unpredictable nature of a signal over various frequency bands is quantified by spectral entropy.

Spectral skewness is a measure of the PSD distribution's asymmetries in a certain frequency band. The ratio of a PSD distribution's peak power to its average power is measured by the spectral crest factor. The mean delta power and lowest delta power are non-EEG characteristics. A brain region's or time period's mean delta power is a measurement of the average power in the delta frequency range. It divulges details regarding the general amount of slow wave activity. The minimal power in the delta frequency range across a specific time period or brain region is referred to as the lowest delta power. The clinical characteristics comprised the Mini-Mental State Examination (MSME) and visual association test (VAT) scores as well as the hallucination (Y/N) score (40). The MSME is a popular cognitive screening instrument that evaluates a variety of cognitive domains: orientation, attention, memory, language, and visuospatial abilities (40). The VAT score evaluates a person's capacity for visual episodic memory by asking them to identify visual stimuli that they have seen in the past (40).

10.3.3 MACHINE LEARNING ALGORITHMS

A diverse set of machine learning algorithms has been extensively employed in dementia research, catering to various data types and research inquiries. The choice of algorithm relies on the specific objectives, available data, and desired outcomes. Some commonly used algorithms for dementia detection encompass support vector machine (SVM), k-nearest neighbors (KNN), random forest (RF), logistic regression (41), decision tree (42), neural network (NN), convolutional neural network (CNN), and recurrent neural network (43). In this part, the primary focus is on utilizing SVM, KNN, RF, NN, and CNN algorithms for dementia detection.

10.3.3.1 Support Vector Machine

SVM is a classification algorithm that aims to differentiate between instances of two classes with maximum accuracy (22). It achieves this by finding a categorization function, represented by a hyperplane, that separates the two classes if the dataset can be linearly divided. The ideal function is found by maximizing the difference between the classes, known as the margin. The margin is the smallest distance between two neighboring instances on the hyperplane. By optimizing the margin, SVM selects the best hyperplanes as solutions.

10.3.3.2 K-nearest Neighbors

KNN is a nonparametric algorithm that compares the training dataset and the study population using various distance measurements to determine similarity. When classifying a new instance, KNN identifies the k-training instances that are most similar to the new instance (27). By using the classifications of these k instances, the algorithm categorizes the new instance based on a majority vote of their classes.

10.3.3.3 Random Forest

RF utilizes a random sampling approach to create a decision tree (44). It selects data randomly, with replacement, from a pool of N samples that is the same size as the initial training set. This means that the same sample can be chosen multiple times. The learning algorithm generates a classifier from each piece of sampled data, and the final classifier is formed by combining all the classifiers from multiple trials. During classification, each classifier votes for the class it belongs to, and the instance is assigned to the class with the most votes. If there is a tie, a random choice is made to determine the winner. Each tree in the ensemble is independently constructed using a bootstrap replica of the input data.

10.3.3.4 Neural Network

The feedforward neural network (FNN) or multilayer perceptron (MLP) is a subclass of artificial neural networks that consists of multiple layers of feedforward-organized neurons (30). Each neuron in the FNN receives input from the layer above, computes the weighted sum of those inputs using a nonlinear activation function, and passes the results to the layer above. The standard architecture of an FNN includes an input layer, one or more hidden layers, and an output layer (26). The input layer corresponds to the input data features, while the output layer represents the expected network output. The hidden layers perform intermediate computations on the input data to extract higher-level characteristics.

Activation functions such as sigmoid, tanh, or ReLU are typically used in the MLP's hidden layers, while the activation function in the output layer depends on the specific problem being solved, commonly SoftMax or sigmoid. To train an FNN or MLP, supervised learning methods like backpropagation are employed to minimize the discrepancy between the predicted output and the actual output (30). This involves iteratively calculating the gradients of the loss function with respect to the weights and biases and updating them in the opposite direction of the gradient.

10.3.3.5   Deep Learning

The convolutional neural network (CNN) is specifically designed for processing data with a grid-like structure, such as images and videos. It consists of different layers, each serving a specific function on the input data. These layers include fully connected, pooling, and convolutional layers, among others. During the convolutional layer, the network applies filters to the input image, allowing it to detect various visual patterns like edges, corners, and textures. These filters are learned during the training process. The pooling layer reduces the spatial dimensions of the feature maps, making the network computationally efficient and more responsive to changes in the input (45).

The final classification decision is made by the fully connected layer, which extracts features from the preceding layers. This layer, also known as the dense layer, is typically followed by one or more fully connected layers. The output of the last pooling layer is flattened into a one-dimensional vector before being fed into the fully connected layers (28). The final classification probabilities are obtained by passing the output of the last fully connected layer through an activation function such as SoftMax. CNNs have been successful in tasks like object detection, face recognition, and scene segmentation in both academic and business applications. They are continually improved through design adjustments and training strategies to enhance performance and generalization capabilities.

10.4   DISCUSSION

10.4.1   ADVANTAGES AND LIMITATIONS OF STATE-OF-THE-ART

Recent research on EEG-based dementia detection led to advances in different aspects. First, more dementia-related features have been proposed in different domains. Advanced signal processing methods have been applied to finding multidimensional, multidomain EEG features. Second, the integration of different machine learning methods, as well as deep learning algorithms have been explored to improve the performance of dementia detection. Different machine learning algorithms were comprehensively compared in some studies (30). Among different types and stages of dementia, there is no single best approach, although deep learning provides a promising tool of reliable diagnosis since it can enable the analysis of different data types and leverage transfer learning which overcomes the limitations of availability of a large neuroimaging data (46). In Table 10.1, the performance metrics are mostly above 0.7, which indicates the applicability of EEG in dementia detection. Future improvement in relevant features, classification algorithms, and large-scale datasets will play a key role in improving the efficacy of EEG-based dementia detection.

There are some limitations in existing works. First, it is important to acknowledge that most previous studies have relied on small datasets (Table 10.1), with the exception of So et al. (47). This reliance on small datasets could compromise the generalizability and reliability of the findings. As a result, the conclusions drawn from these studies may not accurately represent the entire population or capture the full range of variability within the target population. However, it should be noted that the dataset So et al. used (47) was not balanced, and the study did not apply any oversampling techniques. These factors may have introduced biases or limitations in the study's findings.

Second, the existing works lacked in-depth exploration of the underlying mechanisms or pathophysiology related to the phenomenon under investigation. This limitation highlights the need for comprehensive research to enhance our understanding of the biological or physiological processes involved. Third, multiple factors contribute to the current lack of readiness for clinical application in these works.

First, the methods or algorithms proposed in these studies (Table 10.1) did not undergo any external validation processes to ensure their accuracy, reliability, and safety in real-world clinical settings. Second, the developed algorithms or models might be complex and face challenges in integration with existing medical devices or systems, hindering their practical implementation and adoption in clinical practice. Lastly, EEG sensors, which typically require multiple electrodes, may be less convenient to use compared to ECG or photoplethysmogram sensors. This inconvenience can limit their practicality and acceptance in clinical settings.

10.4.2 Future Directions

EEG is cost-effective, accessible, and applicable for the large-scale screening of neurological diseases in underserved populations. EEG readings with fewer scalp electrodes (48), miniaturized design, and dry electrodes will be uniquely useful in the detection of dementia at the pre-symptomatic phase (49). Recently, the Cuban Human Brain Mapping Project repository has been proposed, an open multimodal neuroimaging and cognitive dataset from 282 young and middle-aged healthy participants (50). Low-cost portable EEG devices might lead to the intriguing prospect of the monitoring and evaluation of dementia progression in large populations (Figure 10.3).

To achieve early dementia intervention and personalized health care, fine-grained classification and quantitative evaluation are important approaches that provide detailed references regarding the type and severity. EEG showed efficacy in differentiating the stages of dementia development (36, 51), and identifying different types of early-stage dementia, e.g., vascular dementia (31), frontotemporal dementia and Alzheimer's disease (52). Advanced EEG signal analysis could quantify the disease progression using brain functional connectivity and the identified networks can aid in the diagnosis of dementia-associated disorders (53).

Since neurodegenerative diseases usually present heterogeneous profiles across different levels with different risk factors, compared with unimodal characterization by a single biomarker, AI-assisted analysis of multimodal markers can effectively improve the performance of early dementia detection (54). In a recent study protocol for a cross-sectional trial, a multimodal measurement approach of dementia detection was proposed based on the combination of three non-invasive measurement modalities: EEG, functional near-infrared spectroscopy and heart rate variability derived from electrocardiogram (55). Considering the globally aging population and the increasing incidence of dementia, a multistep process of risk evaluation using a "biomarker pyramid" is recommend for public health-oriented strategies of dementia screening, where low-cost, noninvasive, highly sensitive but nonspecific markers from EEG and other neuroimaging techniques will consist the first step, followed by more expensive/complex markers as (e.g., Aβ and tau titration in cerebrospinal fluid and blood as well as PET with amyloid/tau radioligands (56). Meanwhile, the data from different

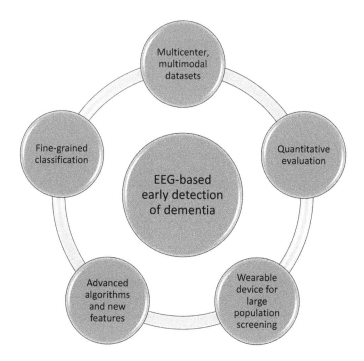

FIGURE 10.3 Future directions of EEG-based detection of dementia.

cohort could validate the identified features/biomarkers from a wider demographic view. Multicenter, multimodal datasets could enable the integration of advanced EEG features with other biomarkers for the more reliable detection of dementia.

REFERENCES

1. Guidelines WHO. Risk Reduction of Cognitive Decline and Dementia. Geneva: World Health Organization. 2019. https://www.who.int/publications/i/item/9789241550543
2. Cantarero-Prieto D, Leon PL, Blazquez-Fernandez C, Juan PS, Cobo CS. The Economic Cost of Dementia: A Systematic Review. Dementia. 2020;19(8):2637–2657.
3. Ontario HQ. The Appropriate Use of Neuroimaging in the Diagnostic Work-Up of Dementia: An Evidence-Based Analysis. Ont Health Technol Assess Ser. 2014;14(1):1–64.
4. Fierini F. Journal of the Neurological Sciences Mixed Dementia: Neglected Clinical Entity or Nosographic Artifice? J Neurol Sci [Internet]. 2020;410(October 2019):116662. Available from: https://doi.org/10.1016/j.jns.2019.116662
5. Canavan M, O'Donnell MJ. Hypertension and Cognitive Impairment: A Review of Mechanisms and Key CHypertension and Cognitive Impairment: A Review of Mechanisms and Key Conceptsoncepts. Front Neurol. 2022;13(February):1–9.
6. Lee S, Zhou J, Wong WT, Liu T, Wu WKK, Wong ICK, et al. Glycemic and Lipid Variability for Predicting Complications and Mortality in Diabetes Mellitus Using Machine Learning. BMC Endocr Disord. 2021;21(1):1–15.
7. Appleton JP, Scutt P, Sprigg N, Bath PM. Hypercholesterolaemia and Vascular Dementia. Clin Sci. 2017;131(14):1561–1578.
8. Livingston G, Huntley J, Sommerlad A, Ames D, Ballard C, Banerjee S, et al. Dementia Prevention, Intervention, and Care: 2020 Report of the Lancet Commission. Lancet. 2020;396(10248):413–446.

9. Nordestgaard LT, Christoffersen M. Shared Risk Factors between Dementia and Atherosclerotic Cardiovascular Disease. Int J Mol Sci. 2022;23(17):9777.

10. Sindi S, Mangialasche F, Kivipelto M. Advances in the Prevention of Alzheimer's Disease. Fprime Reports. 2015;7:1–12.

11. Zhou J, Lee S, Wong WT, Waleed KB, Leung KSK, Lee TTL, et al. Gender-Specific Clinical Risk Scores Incorporating Blood Pressure Variability for Predicting Incident Dementia. J Am Med Informatics Assoc. 2022;29(2):335–347.

12. Lee S, Zhou J, Liu, T, et al. Gender-Specific Clinical Risk Scores Incorporating Blood Pressure Variability for Predicting Incident Dementia. J Am Med Informatics Assoc. 2022;29(10):1825–1826.

13. Ahmed R, Zhang Y, Member S, Feng Z, Lo B, Inan OT, et al. Neuroimaging and Machine Learning for Dementia Diagnosis: Recent Advancements and Future Prospects. IEEE Rev Biomed Eng. 2018;12:19–33.

14. Maki Y, Yamaguchi H. Early Detection of Dementia in the Community Under a Community-Based Integrated Care System. Geriatr Gerontol Int. 2014;14(S2):2–10.

15. De Roeck EE, De Deyn PP, Dierckx E, Engelborghs S. Brief Cognitive Screening Instruments for Early Detection of Alzheimer's Disease: A Systematic Review. Alzheimer's Res Ther. 2019;11(1):1–14.

16. Burns S, Selman A, Sehar U, Rawat P, Reddy AP, Reddy PH. Therapeutics of Alzheimer's Disease: Recent Developments. Antioxidants (Basel). 2022;11(12):2402.

17. Rodríguez-reséndiz J, Member S, Avecilla-ramírez GN. Impact of EEG Parameters Detecting Dementia Diseases: A Systematic Review. IEEE Access. 2021;9:78060–78074.

18. Hosseini M, Member S, Hosseini A, Review M. A Review on Machine Learning for EEG Signal Processing in Bioengineering. IEEE Rev Biomed Eng. 2021;14:204–218.

19. Battineni G, Chintalapudi N, Hossain MA, Losco G, Ruocco C, Sagaro GG, et al. Artificial Intelligence Models in the Diagnosis of Adult-Onset Dementia Disorders: A Review. Bioengineering. 2022;9(8).

20. Anastasio TJ. Predicting the Potency of Anti-Alzheimer's Drug Combinations Using Machine Learning. Processes. 2021;9(2)(264).

21. Tsoi KKF, Jia P, Dowling N, Titiner JR, Wagner M, Capuano A, Donohue MC. Applications of Artificial Intelligence in Dementia Research. Cambridge Prism Precis Med. 2023;1(e9):1–9.

22. De La Fuente Garcia S, Ritchie CW, Luz S. Artificial Intelligence, Speech, and Language Processing Approaches to Monitoring Alzheimer's Disease: A Systematic Review. J Alzheimer's Dis. 2020;78(4):1547–1574.

23. Ebrahimighahnavieh MA, Luo S, Chiong R. Deep Learning to Detect Alzheimer's Disease from Neuroimaging: A Systematic Literature Review. Comput Methods Programs Biomed [Internet]. 2020;187:105242. Available from: https://doi.org/10.1016/j.cmpb.2019.105242

24. Suk H-I, Lee S-W, Shen D. Deep Sparse Multi-Task Learning for Feature Selection in Alzheimer's Disease Diagnosis. Physiology & Behavior. 2016;176:1–15.

25. Jain V, Nankar O, Jerrish DJ, Gite S, Patil S, Kotecha K. A Novel AI-Based System for Detection and Severity Prediction of Dementia Using MRI. IEEE Access. 2021;9: 154324–154346.

26. Li R, Wang X, Lawler K, Garg S, Bai Q, Alty J. Applications of Artificial Intelligence to Aid Early Detection of Dementia: A Scoping Review on Current Capabilities and Future Directions. J Biomed Inform [Internet]. 2022;127(January):104030. Available from: https://doi.org/10.1016/j.jbi.2022.104030

27. Durongbhan P, Zhao Y, Chen L, Zis P, De Marco M, Unwin ZC, et al. A Dementia Classification Framework Using Frequency and Time-Frequency Features Based on EEG Signals. IEEE Trans Neural Syst Rehabil Eng. 2019;27(5):826–835.

28. Sharma N, Kolekar MH, Jha K, Kumar Y. EEG and Cognitive Biomarkers Based Mild Cognitive Impairment Diagnosis. Irbm [Internet]. 2019;40(2):113–121. Available from: https://doi.org/10.1016/j.irbm.2018.11.007

29. Sharma N, Kolekar MH, Jha K. EEG Based Dementia Diagnosis Using Multi-Class Support Vector Machine with Motor Speed Cognitive Test. Biomed Signal Process Control [Internet]. 2021;63:102102. Available from: https://doi.org/10.1016/j.bspc.2020.102102

30. Ieracitano C, Mammone N, Hussain A, Morabito FC. A Novel Multi-Modal Machine Learning Based Approach for Automatic Classification of EEG Recordings in Dementia. Neural Networks [Internet]. 2020;123:176–190. Available from: https://doi.org/10.1016/j.neunet.2019.12.006

31. Wang C, Xu J, Zhao S, Lou W. Identification of Early Vascular Dementia Patients with EEG Signal. IEEE Access. 2019;7:68618–68627.

32. Anuradha G, Jamal DN. Classification of Dementia in EEG with a Two-Layered Feed Forward Artificial Neural Network. Eng Technol Appl Sci Res. 2021;11(3):7135–7139.

33. Jennings JL, Peraza LR, Baker M, Alter K, Taylor JP, Bauer R. Investigating the Power of Eyes Open Resting State EEG for Assisting in Dementia Diagnosis. Alzheimer's Res Ther [Internet]. 2022;14(1):1–12. Available from: https://doi.org/10.1186/s13195-022-01046-z

34. Li F, Matsumori S, Egawa N, Yoshimoto S, Yamashiro K, Mizutani H, et al. Predictive Diagnostic Approach to Dementia and Dementia Subtypes Using Wireless and Mobile Electroencephalography: A Pilot Study. Bioelectricity. 2022;4(1):3–11.

35. Li Y, Xiao S, Li Y, Li Y, Yang B. Classification of Mild Cognitive Impairment from Multi-Domain Features of Resting-State EEG. In: Annual International Conference of the IEEE Engineering in Medicine & Biology Society (EMBC). 2020.

36. Ieracitano C, Mammone N, Bramanti A, Hussain A, Morabito FC. A Convolutional Neural Network Approach for classification of Dementia Stages Based on 2D-Spectral Representation of EEG Recordings. Neurocomputing [Internet]. 2019;323:96–107. Available from: https://doi.org/10.1016/j.neucom.2018.09.071

37. Al-qazzaz NK, Hamid S, Ali B, Ahmad SA, Chellappan K, Islam S, et al. Role of EEG as Biomarker in the Early Detection and Classification of Dementia. Sci World J. 2014;(906038).

38. Sharma N, Member S, Kolekar MH, Member S, Jha K. Iterative Filtering Decomposition based Early Dementia Diagnosis using EEG with Cognitive Tests. IEEE Trans Neural Syst Rehabil Eng. 2020;28(9):1890–1898.

39. Brookshire G, Merrill DA, Wu YC, Caselli RJ, Quirk C, Gerrol S, et al. Alzheimer's Disease Status Can be Predicted Using a Novel Fractal-based Metric Computed from Resting-State EEG. Alzheimer's Demensia. 2022;18:e067509.

40. Dauwan M, van der Zande JJ, van Dellen E, Sommer IEC, Scheltens P, Lemstra AW, et al. Random Forest to Differentiate Dementia with Lewy Bodies from Alzheimer's Disease. Alzheimer's Dement: Diagn, Assess Dis Monit. 2016;4:99–106.

41. Mazzocco T, Hussain A. Novel Logistic Regression Models to Aid the Diagnosis of Dementia. Expert Syst Appl [Internet]. 2012;39(3):3356–3361. Available from: http://dx.doi.org/10.1016/j.eswa.2011.09.023

42. Al-Dlaeen D, Alashqur A. Using Decision Tree Classification to Assist in the Prediction of Alzheimer's Disease. In: 2014 6th International Conference on Computer Science and Information Technology, CSIT 2014—Proceedings. IEEE; 2014. pp. 122–126.

43. Nori VS, Hane CA, Sun Y, Crown WH, Bleicher PA. Deep Neural Network Models for Identifying Incident Dementia Using Claims and EHR Datasets. PLoS One [Internet]. 2020;15(9 September):1–12. Available from: http://dx.doi.org/10.1371/journal.pone.0236400

44. Rye I, Vik A, Kocinski M, Lundervold AS, Lundervold AJ. Predicting Conversion to Alzheimer's Disease in Individuals with Mild Cognitive Impairment Using Clinically Transferable Features. Sci Rep [Internet]. 2022;12(1):15566. Available from: https://doi.org/10.1038/s41598-022-18805-5

45. Xia W, Zhang R, Zhang X, Usman M. Heliyon A Novel Method for Diagnosing Alzheimer's Disease Using Deep Pyramid CNN based on EEG Signals. Heliyon [Internet]. 2023;9(4):e14858. Available from: https://doi.org/10.1016/j.heliyon.2023.e14858

46. Mirzaei G, Adeli H. Machine Learning Techniques for Diagnosis of Alzheimer Disease, Mild Cognitive Disorder, and Other Types of Dementia. Biomed Signal Process Control [Internet]. 2022;72:103293. Available from: https://doi.org/10.1016/j.bspc.2021.103293

47. So A, Hooshyar D, Park KW, Lim HS. Early Diagnosis of Dementia from Clinical Data by Machine Learning Techniques. Appl Sci. 2017;7(7).

48. Del Percio C, Lopez S, Noce G, Lizio R, Tucci F, Soricelli A, et al. What a Single Electroencephalographic (EEG) Channel Can Tell us About Alzheimer's Disease Patients with Mild Cognitive Impairment. Clin EEG Neurosci. 2023;54(1):21–35.

49. Koenig T, Smailovic U, Jelic V. Psychiatry Research: Neuroimaging Past, Present and Future EEG in the Clinical Workup of Dementias. Psychiatry Res Neuroimaging [Internet]. 2020;306:111182. Available from: https://doi.org/10.1016/j.pscychresns.2020.111182

50. Valdes-sosa PA, Galan-garcia L, Bosch-bayard J, Bringas-vega ML, Aubert-vazquez E, Rodriguez-gil I, et al. The Cuban Human Brain Mapping Project, a Young and Middle Age Population-based EEG, MRI, and Cognition Dataset. Sci Data [Internet]. 2021;8(1):45. Available from: http://dx.doi.org/10.1038/s41597-021-00829-7

51. Amezquita-sanchez JP, Mammone N, Morabito FC, Adeli H. A New Dispersion Entropy and Fuzzy Logic System Methodology for Automated Classification of Dementia Stages Using Electroencephalograms. Clin Neurol Neurosurg [Internet]. 2021;201:106446. Available from: https://doi.org/10.1016/j.clineuro.2020.106446

52. Miltiadous A, Tzimourta KD, Giannakeas N, Tsipouras MG, Afrantou T, Ioannidis P, et al. Alzheimer's Disease and Frontotemporal Dementia: A Robust Classification Method of EEG Signals and a Comparison of Validation Methods. Diagnostics. 2021;11(8).

53. Adebisi AT, Gonuguntla V, Lee H, Veluvolu KC. Classification of Dementia Associated Disorders Using EEG based Frequent Subgraph Technique. In: 2020 International Conference on Data Mining Workshops (ICDMW). 2020. pp. 613–620.

54. Moguilner S, Birba A, Fittipaldi S, Gonzalez-Campo C, Tagliazucchi E, Reyes P, et al. Multi-Feature Computational Framework for Combined Signatures of Dementia in Underrepresented Settings. J Neural Eng. 2022;19(4):046048.

55. Grässler B, Herold F, Dordevic M, Gujar TA, Darius S, Böckelmann I, et al. Multimodal Measurement Approach to Identify Individuals with Mild Cognitive Impairment: Study Protocol for a Cross- Sectional Trial. BMJ Open. 2021;11(5):e046879.

56. Rossini PM, Miraglia F, Vecchio F. Early Dementia Diagnosis, MCI-to-Dementia Risk Prediction, and the Role of Machine Learning Methods for Feature Extraction from Integrated Biomarkers, in Particular for EEG Signal Analysis. Alzheimer& Dement. 2022;18(12):2699–2706.

57. Trambaiolli LR, Lorena AC, Fraga FJ, Kanda PAM, Anghinah R, Nitrini R. Improving Alzheimer's Disease Diagnosis with Machine Learning Techniques. Clin EEG Neurosci. 2011;42(3):160–165.

58. Ieracitano C, Mammone N, Morabito FC. A Convolutional Neural Network based Self-Learning Approach for Classifying Neurodegenerative States from EEG Signals in Dementia. In: 2020 International Joint Conference on Neural Networks (IJCNN). 2020.

59. Amini M, Pedram MM, Moradi AR, Ouchani M. Diagnosis of Alzheimer's Disease by Time-Dependent Power Spectrum Descriptors and Convolutional Neural Network Using EEG Signal. Comput Math Methods Med. 2021;2021:5511922.

11 EEG Signal Processing-Driven Machine Learning for Cognitive Task Recognition

Siran Wang, Brian Lee, Gary Tse, and Haipeng Liu

11.1 INTRODUCTION

Cognition refers to the mental processes involved in acquiring knowledge and understanding through thought, experience, and the senses (1). A cognitive task is a group of related mental activities directed toward a goal that may not always be clear (2). These tasks can range in complexity from simple ones like identifying colors or shapes to more intricate ones like solving mathematical equations, reading comprehension, or decision-making. As a general concept, cognitive tasks include undertakings that the human mind can think of or perform, such as a color, feeling, image, or text (3). Emotional changes can also significantly influence the subject's response in these tasks (4). While cognitive task activities are not fully observable (2), they are often used to identify brain networks involved in underlying cognitive processes. To directly observe and quantitatively analyze the cognitive task activities, different neuroimaging and electrophysiological techniques have been developed, e.g., functional magnetic resonance imaging (fMRI), functional near-infrared spectroscopy, magnetoencephalography (MEG), and electroencephalography (EEG).

EEG is a neuroimaging technique used to study brain activity by measuring the electric potential on the scalp (5). It is particularly useful for detecting neuronal activations and their interactions at the macroscopic level, as these are primarily of electromagnetic origin and can be recorded in the form of EEG signals with a resolution of milliseconds. EEG signals are generated by the average of intracellular and extracellular ionic currents produced by masses of synchronized neurons that are active in a particular area of the brain at the same time (6).

Figure 11.1 shows the workflow of EEG signal processing. EEG is a noninvasive technique that boasts high time-resolution and broad applicability (7). Compared with other brain bio-signals, EEG is more cost-efficient and provides a wider range of information regarding the mental state of a subject (8, 9), making it a promising technology in cognitive task recognition. Furthermore, EEG has been established as a versatile tool in various fields, including brain–computer interface (BCI), the

DOI: 10.1201/9781003479970-11

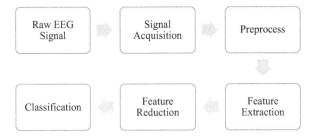

FIGURE 11.1   Flow diagram of processing of EEG signals for cognitive task recognition.

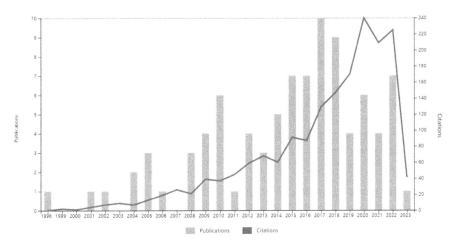

FIGURE 11.2   Number of studies and citation frequency.

diagnosis of neurological and neurosurgical diseases like epilepsy, head injury, and brain tumors (9), as well as stress and cognitive task-load (CTL) detection.

However, accurately characterizing brain activity during a cognitive task is challenging due to the complex nature of the brain (10). Brain activities are highly dynamic, irregular, transient, and non-stationary, especially those related to cognitive and behavioral events (11). Additionally, EEG data are high dimensional, sparse, and noisy, making neural data processing an even more daunting task (12). Results of cognitive task recognition using EEG highly depend on the choice of processing methods (13).

Therefore, improving cognitive task recognition using EEG requires addressing various challenges, such as removing motion artifacts in EEG data, eliminating muscle contamination, and improving the robustness of brain signal decoding against individual difference. With recent development in neuroscience, there is an increasing need to extract relevant information from EEG recordings to observe the desired cognitive processes and understand the underlying neural mechanisms, whereas the large size of EEG data makes visual analysis difficult if not impossible (14).

In the past two decades, a diverse range of methods has been proposed to address the challenges in EEG-based cognitive task recognition, suggested by increasing number of studies and citations in Figure 11.2. Artificial intelligence (AI) methods, especially machine learning algorithms, have reshaped the approaches of EEG signal processing and its applications. New algorithms have been proposed to suit different cognitive tasks and application scenarios. However, there is a scarcity of comprehensive review in this area. This chapter aims to provide an overview of the existing literature on the applications of machine learning in EEG-based cognitive task recognition, clarify its significance and limitations, and identify new research directions.

11.2 LITERATURE SEARCH STRATEGY

This systematic review was conducted in accordance with the Preferred Reporting Items for Systematic Review and Meta-analysis (PRISMA) guidelines and its methodology.

11.2.1 SEARCH STRATEGY

Web of Science (WoS) was searched for articles published between January 2010 and February 2023. The search in WoS used the following terms: TS = ("cognitive task") AND TS = ("recognition") AND (TS = ("electroencephalogram") OR TS = ("EEG") OR TS = ("electroencephalograph")). The articles were then assessed for their eligibility for selection.

11.2.2 INCLUSION CRITERIA

The inclusion criteria were studies that 1) involved cognitive tasks, 2) used EEG recording during the cognitive task recognition experiment, and 3) included AI algorithms. Articles with simple statistical analyses or irrelevant results were excluded.

11.2.3 STUDY SELECTION

The articles extracted from the database were managed using the reference management software package Endnote, and an Excel spreadsheet was used for the subsequent eligibility analysis. Two reviewers independently evaluated the articles by checking their titles, abstracts, or the entire document if necessary. In cases where there was not agreement between the reviewers, the article's inclusion was resolved through group discussion. The relevant information for the qualitative analysis of the articles included in the review was extracted using the Review Manager (RevMan), which provides guidelines for writing various types of systematic reviews. This software also allows for the addition, ordering, and reference of studies, as well as the generation of PRISMA flow diagrams and tables, data classification, and analysis.

11.3 RESULTS

11.3.1 TYPES OF COGNITIVE TASKS

As summarized in Table 11.1, in the reviewed studies, various types of cognitive tasks were used, with most studies focusing on one or a few specific tasks. The type

TABLE 11.1
The Existing Methods for Cognitive Task Recognition Using EEG Signals

Title	Author,year	Subject (number, age)	Data type	Dataset	Cognitive task	Algorithm	Performance	Validation
Design of an Adaptive Human-Machine System Based on Dynamical Pattern Recognition of Cognitive Task-Load	Jianhua Zhang et al., 2017 (18)	7 (22–24)	EEG data, ECG data, EOG data	from participants	simulated process control tasks	Preprocess: adaptive exponential smoothing Feature deduction: locality preserving projection Classification: least-square support vector machine (SVM)	An overall correct classification rate of about 80%	/
Classification of EEG signals in an Object Recognition task	Rus et al., 2017 (22)	4, /	EEG data	from participants	an object recognition task	Feature extraction: fast Fourier transform (FFT) Classification: SVM, K-nearest neighbors, and artificial neural networks	Best accuracy (ACC): 95.1% for a binary classification 87% for a multiclass classification	/

(Continued)

TABLE 11.1 (Continued)
The Existing Methods for Cognitive Task Recognition Using EEG Signals

Title	Author,year	Subject (number, age)	Data type	Dataset	Cognitive task	Algorithm	Performance	Validation
Information-Theoretic Measures on Intrinsic Mode Function for the Individual Identification Using EEG Sensors	Kumari et al., 2015 (23)	7, /	EEG data	from participants	Five sets of mental tasks (breathing, mental mathematics, geometric figure rotation, visual counting and mental letter composing, etc.)	Preprocess: Empirical mode decomposition (EMD) Feature deduction: locality preserving projection; Feature extraction: EMD, intrinsic mode function Classification: learning vector quantization neural network with fuzzy neuro algorithm	best recognition rate: over 97%	10-fold cross validation
Enhancing the Classification of EEG Signals using Wasserstein Generative Adversarial Networks	Petruțiu et al., 2020 (24)	N/A	EEG data	from participants	a human visual recognition task	Preprocess: filtering, Principal component analysis (PCA); Generation of artificial EEG signals:	Accuracy: significant improvement in the classification	leave-one-out cross-validation (LOOCV)

Title	Author, Year	Participants	Data	Task	Methods	Results	Validation
Multimodal Approach for Cognitive Task Performance Prediction from Body Postures, Facial Expressions and EEG Signal	Ashwin Ramesh Babu et al., 2018 (1)	15, 22–35	behavior data such as Body Postures, Facial Expressions, EEG data from participants, the Kaggle facial expression recognition (FER) challenge dataset 2013	the Sequence Learning Task	Generative adversarial networks Classification: Convolutional neural network SVM, gradient boosting, random forest (RF), embedded training, edited nearest neighbor, algorithm 2: Final performance prediction combined from three modalities	highest accuracy: 87.5 percent	10-fold cross validation
Frontal EEG-Based Multi-Level Attention States Recognition Using Dynamical Complexity and Extreme Gradient Boosting	Wan et al., 2021 (7)	42, 20–26	EEG data, basic information such as handiness' and sleep quality from participants	a sustained attention task	Preprocess: Independent component correlation Feature reduction: PCA Classification: Complexity-XGBoost (SVM, and RF)	Maximum average accuracy: 81.39 ± 1.47% for four attention levels, 80.42 ± 0.84% for three attention levels, and 95.36 ± 2.31% for two attention levels	LOOCV and 5-fold-cross validation

(Continued)

TABLE 11.1 (Continued)
The Existing Methods for Cognitive Task Recognition Using EEG Signals

Title	Author, year	Subject (number, age)	Data type	Dataset	Cognitive task	Algorithm	Performance	Validation
Pattern Classification of Instantaneous Cognitive Task-load Through GMM Clustering, Laplacian Eigenmap, and Ensemble SVMs	Donos et al., 2022 (19)	7, 22–24	EEG data, heart rates	from participants	the control task provided by automation enhanced Cabin Air Management System	Feature extraction: Laplacian eigenmap (LE) Feature reduction: LE Classification: a Gaussian mixture model, multiple SVMs	subject-average ACC 0.7677	the standard cross-validation technique is not employed here
Detecting Autism Spectrum Disorder Using Topological Data Analysis	Majumder et al., 2021 (12)	12 autistic and 12 typically developing, 6–13	EEG data	from participants	a visual cognitive task	Feature extraction: probabilistic data association	ACC: 90%	10-fold stratified cross-validation
Thalamic bursts modulate cortical synchrony locally to switch between states of global functional connectivity in a cognitive task	Portoles et al., 2022 (21)	18	EEG data, Magnetoencephalography data	from participants	a memory task	an optimization algorithm in combination with hidden semi-Markov models to divide the cognitive task into stages with separate connectivity patterns	unveil the dynamics of local and cortex-wide neural coordination underlying the fundamental cognitive processes involved in a memory task.	leave-one-subject-out cross-validation

								LOOCV
EEG Pattern Recognition using Brain-Inspired Spiking Neural Networks for Modelling Human Decision Processes	Doborjeh, ZG et al., 2018 (25)	23, average 27	EEG data	from participants	a moral dilemma situation-related task	Preprocess: the learning algorithm Classification: the dynamic evolving spiking neural network algorithm	total ACC: 85%	
EEG-based measurement system for monitoring student engagement in learning 4.0	Apicella et al., 2022 (26)	21, 23.7 average	EEG data	from participants	an experimental task to learn how a specific human–machine interface works	Feature extraction: filter bank, common spatial pattern Classification: SVM	average accuracy: almost 77%	/
Improved artefact removal from EEG using Canonical Correlation Analysis and spectral slope	Janani, AS et al., 2018 (27)	6, 13, 93	EEG data	from participants	a series of tasks including: baseline eyes closed, baseline eyes open, the auditory verbal learning task, etc	Preprocess: a new automatic muscle-removal approach based on the traditional blind source separation-canonical correlation analysis and the spectral slope of its components	significantly lower computational complexity	/

(Continued)

TABLE 11.1 (*Continued*)
The Existing Methods for Cognitive Task Recognition Using EEG Signals

Title	Author, year	Subject (number, age)	Data type	Dataset	Cognitive task	Algorithm	Performance	Validation
Cognitive Task classification using Fuzzy based Empirical Wavelet Transform	Tanveer, M et al., 2019 (17)	6	EEG data	publicly available EEG database for the mental task classification (45).	a mental task	Classification based on fuzzy c-means algorithm	higher average classification accuracy fewer features	/
Evolving dynamic clustering of spatial/ Spectro-temporal data in 3D spiking neural network models and a case study on EEG data	Doborjeh et al., 2018 (28)	3 groups	EEG data	from participants	a cognitive GO-NOGO task	Classification: clustering and the NeuCube spiking neural network architecture one performed during unsupervised and one during supervised learning models.	higher clustering accuracy	/
Classification of left and right foot kinaesthetic motor imagery using common spatial pattern	Tariq, M et al., 2019 (20)	9, 21–28	EEG data	from participants	foot motor tasks	Classification: CSP linear discriminant analysis (LDA), CSP Logreg, and filter bank common spatial pattern	maximum accuracy: 77.5%	10-fold cross validation

| Auditory and spatial navigation imagery in Brain–computer Interface using optimized wavelets | Cabrera, AF et al., 2008 (29) | 19, 21–24 | EEG data | from participants | Two non-motor imagery tasks | Feature extraction: / adaptive replacement cache, frame rate control, overlap weighting (OW) 4, OW6, and discrete wavelet transform (DWT) using mother wavelets from the Daubechies family Classification: a Bayesian classifier. Apriori, DWT using Daubechies wavelets and DWT using adaptive wavelets) | 5-fold cross-validation |

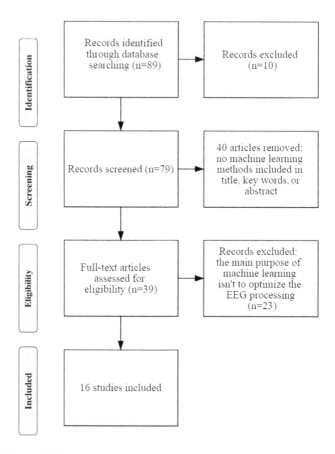

FIGURE 11.3 PRISMA flow diagram of the selection of the studies for this review.

of cognitive task used has a significant impact on the effectiveness of EEG-based recognition. For instance, motor imagery is the most commonly used task in BCI studies which generates detectable changes in EEG during movement planning (15).

11.3.2 DATA

11.3.2.1 Data Types

In addition to EEG data, some studies also utilized clinical imaging data (e.g. fMRI) and medical records where demographic information, medical history, and lab test results are included. These additional data types played a crucial role in validating the conclusions (16). In some studies, fMRI data served for comparison with EEG data, while basic information helped eliminate interference and obtain more accurate data and concise results (6).

11.3.2.2 Datasets

Datasets are a crucial part of machine learning (Figure 11.4). The utility of different datasets has a great impact on the reliability of the algorithms. A well-curated and

FIGURE 11.4   Word cloud map of articles.

diverse EEG dataset can help the machine learning algorithm to learn more effectively and generalize well to unseen data.

Two studies use public datasets (1, 17). Keirn and Aunon in 1990 used publicly available EEG database for mental task classification in a study of cognitive task classification using fuzzy-based empirical wavelet transform (17). All datasets are BCI datasets where EEG signals are commonly utilized to characterize the brain activities.

In other studies, the EEG datasets were collected from participants. Most of them are healthy subjects, but one study focused on detecting schizophrenia (9) and one focused on autism spectrum disorder (ASD) or patients diagnosed with schizophrenia (12). To ensure the quality of the datasets, data cleansing is performed by excluding the data entries with low-quality EEG recordings or missing information. The number of participants is often below 10. There is a need for large-scale datasets for full validation and better applicability of the conclusions to different cohorts (1). Most participants are under 30 years old. To widen the applicability, people of different age groups should be considered, since the reaction time, which sometimes serves as a measurement, depends significantly on age (7).

11.3.3   Algorithm

11.3.3.1   Overview of Algorithms

Machine learning has emerged as an indispensable tool in EEG signal analysis, providing excellent accuracy and efficiency compared with traditional signal processing techniques. This has enabled researchers to gain deeper insights into cognitive tasks. Support vector machines (SVM) and random forests (RF) are examples of traditional machine learning algorithms commonly used in EEG analysis. Some researchers used existing new algorithms such as least-square SVM (LSSVM) to get better accuracy (18). Additionally, some researchers have proposed new algorithms based on traditional ones. For instance, the extreme gradient boosting (XGBoost) classifier was proposed as the classification model, Complexity-XGBoost, to distinguish multilevel attention states with improved classification accuracy (7).

In conventional approaches, ineffective decoding of EEG features and high complexity of algorithms often lead to unsatisfactory performance. Based on the better classification performance, new machine learning algorithms are proposed focusing on the features, scalability, extensibility, and applicability of the algorithm, e.g., deriving a reasonable number of target classes and low-dimensional optimal EEG features for individual human operator subjects (19). Algorithms targeting on problem-solving are also proposed, such as the classification of left and right foot kinesthetic motor imagery for BCI applications using common spatial patterns (20).

Machine learning algorithms can also be utilized in testing scientific hypothesis. For example, Portoles et al. developed a machine learning approach by combining multivariate pattern analysis with a hidden semi-Markov model that could identify different cognitive stages, to test the hypothesis that thalamic bursts modulate cortical synchrony locally to switch between states of global functional connectivity in a cognitive task (21).

11.3.3.2 Types of Algorithms

11.3.3.2.1 Preprocessing of EEG Signals

In EEG analysis, preprocessing is a critical step that involves cleaning, filtering, and transforming the raw EEG signal prior to further analysis. Some common preprocessing techniques used in EEG analysis include artifact rejection, baseline correction, and filtering. Artifact rejection involves removing segments of the EEG signal that are contaminated by noise or other unwanted signals, such as eye blinks or muscle activity. Baseline correction refers the adjustment of the signal so that the average value is zero, which can help to reduce baseline drifts and other unwanted low-frequency artifacts. Filtering involves removing unwanted frequency components from the signal, such as high-frequency noise or low-frequency drift.

Advanced algorithms can be used to automate some steps and improve the efficiency of EEG preprocessing (30). For example, both independent component analysis (ICA) and canonical correlation analysis (CCA) are blind source separation (BSS) approaches that have been commonly used for removing muscular artifact interference. A scoping review showed that ICA was used in more than one third of recent studies on EEG denoising (31).

ICA is a linear transformation that separates a set of signals into a set of sources or components and can be used to automatically separate artifacts from the EEG signal that overlap in frequency ranges (32). CCA can find the correlation structure between two multivariate datasets and identify the uncorrelated components (31). While both techniques have shown promise in detecting and eliminating signal components associated with muscle noise, they have their disadvantages, such as computational complexity, the need for many samples of data, and the inability to separate sources with Gaussian distributions or white spectra. The standard approach to muscle artifact reduction using CCA has inherent limitations. These include the escalation of spectral power in muscle noise components with increasing frequency, insufficient guidance on component selection for removal, oversight of the impact of environmental noises, and a lack of attention to the effectiveness of the approach in addressing tonic muscular artifact interference. These limitations highlight the need for continued research and development in the field of electromyogram (EMG)-free EEG (33).

In addition to the BSS methods, adaptive signal enhancement (ASE) was shown to be capable of removing motion artifacts with no need of the motion template while simultaneously preserving crucial information about CTL variations. Adaptive data smoothing is also proposed to remove motion artifacts in EEG data (18). Other novel methods can be found in some comprehensive review papers (31, 34, 35). The selection of preprocessing method depends on the aim and application scenario (e.g., clinical diagnosis, BCI, neuromarketing) (35). Some issues, e.g., single- or multi-channel EEG signals, online or real-time application, and reference sensors of motion artefacts (often accelerometers and gyroscopes) also need to be considered in selecting different preprocessing algorithms (36). Considering the heterogeneity of different noises, it is recommended that different algorithms be combined to correct the signal using multiple processing stages (37).

11.3.3.2.2 Feature Extraction

In EEG analysis, feature extraction is fundamental for identifying relevant patterns and events within the data. Feature extraction involves transforming the preprocessed EEG signal into a set of features that capture specific characteristics of the signal in different domains, such as frequency content, amplitude, temporal dynamics, and statistical features (38, 39). Common techniques for feature extraction in EEG analysis include Fourier transform, wavelet transform, time–frequency analysis, and spectral coherence analysis.

Regarding cognitive task recognition, some new algorithms have been applied in recent studies. Empirical mode decomposition (EMD) has been employed to decompose EEG signals into intrinsic mode functions (IMFs) for feature extraction (40). The low-frequency IMFs can be discarded since they do not contain essential information. Laplacian eigenmaps (LE) is a popular manifold learning technique that has been applied to extract EEG spectral markers of variation in CTL levels (19). Topological data analysis (TDA) has also been successfully used for high dimensional data processing, including time-series data analysis and epilepsy detection using EEG signals. TDA can extract important features from data based on the topological structure of the data, which has coordinated independence and deformation invariance (12). A feature extraction pipeline based on a filter bank and common spatial pattern (CSP) has been adopted for investigating the EEG frequency spectrum and improving data separability. The pipeline employs 12 infinite impulse response band-pass Chebyshev type 2 filters and CSP for feature extraction. The combination of filter bank and CSP allows for investigating different frequency intervals and has been effective in mental state detection (21).

11.3.3.2.3   Feature Deduction and Selection

Feature deduction algorithms play a crucial role in identifying the most informative EEG features or channels that contribute to the classification of brain states or events. These algorithms aim to reduce the dimensionality of the EEG data while retaining the most relevant information. This is important because EEG signals contain a large number of channels and time points, which can make the data analysis computationally expensive and prone to overfitting. Feature deduction algorithms use various techniques, such as ICA, principal component analysis, and CSP, to extract the most informative features from the EEG data. For instance, in Zhang et al.'s study, initially there were 56 EEG and ECG related features, but using the locality preserving projection (LPP) technique, this was reduced to a more manageable set of 12 salient features, with 11 of them being related to EEG (18).

11.3.3.2.4   Classification

Classification is the main aim of machine learning algorithms in EEG analysis (Figure 11.4), which identifies specific brain states or activities based on the EEG features. Machine learning algorithms can be trained on labeled EEG datasets to automatically classify new EEG data. Different classification algorithms that can be applied to EEG data, including decision trees, SVM, neural networks, and Bayesian classifiers (27). The choice of algorithm depends on the specific research question and the nature of the data being analyzed (22).

SVM is a conventional binary classifier that has been widely used in various fields. The basic idea of SVM is to find a hyperplane that maximizes the margin between two classes. SVM achieves this by representing the data as points in a vector space and then finding the hyperplane that maximally separates the two classes. This approach is different from other hyperplane-based classifiers, as SVM finds the hyperplane that has the largest distance from the margins of the classes.

Based on SVM, Zhang et al. developed participant-specific dynamic LSSVM model to classify the instantaneous CTL into five classes at each time instant. The novelty of this work lies in the construction of a dynamic model with output feedback and the use of an adaptive learning algorithm to find the optimal parameters of the model (18). The SVM algorithm was used as a multiclass classification algorithm, and an ensemble of independent member classifiers was built to make a final classification decision based on the outputs of individual classifiers. This approach helps to overcome the difficulty of determining the optimal functions for each subject. Additionally, to build an ensemble SVM classifier with high generalizability for predicting CTL levels, different kernel functions and multiclass SVM techniques are integrated to enhance the independence among the member classifiers (19).

In addition to SVM, other algorithms are also used in classification. Hybrid learning for classification is mentioned where learning vector quantization neural network with fuzzy algorithm was used (23). A data augmentation method using generative adversarial networks to generate artificial EEG signals from existing data in order to improve classification performance. The preliminary results suggest that the introduction of artificially generated signals have a positive effect on the performance of the classifier (24). Advanced classification methods, e.g., fuzzy logic, deep learning, ensemble learning, reinforcement learning, and different combinations, are recommended for EEG analysis and deserve exploration in future research on cognitive task recognition (27, 41).

11.3.4 VALIDATION

Validation serves the purpose of testing a model's ability to generalize to new data and assess its overall performance. N-fold cross-validation (n is often set as 5 or 10) based on randomized separation of the dataset is commonly used to estimate the optimal parameters for the classifiers and avoid overfitting classifiers to the training data (8, 16). In 10-fold cross-validation, the system was trained with 90% of the data and tested with 10% of the data. The cross-validation process was performed 10 times with different validation sets to check for consistency of the model across all samples. By changing the value of n, n-fold validation can evaluate the performance of the classification algorithm in different training set sizes (10).

Leave-one-out cross-validation (LOOCV) works by iteratively removing a single observation from the dataset, training the model on the remaining observations, and then testing the model's performance on the observation that was left out. This process is repeated for every observation in the dataset, and the results are aggregated to provide an estimate of the model's performance (**7, 21, 24, 25**). Rizkallah et al. did not use cross validation because there was only a small amount of data at very-high CTL level (i.e., task-load conditions #4 and #5 in a session) (19). For cross validation, the division of dataset based on task-load conditions or simple randomization may lead to a lack of very-high CTL data in the training or testing dataset and hence false classification results. There are also studies that do not mention any information about validation.

11.3.5 EVALUATIONS OF PERFORMANCE

Several evaluation metrics are used to test the effect of the algorithms, such as accuracy, precision, recall, F1-Score, receiver operating characteristic (ROC) curve, and area under ROC curve. For example, the high accuracy (90%) of the pattern classifier validates the efficacy of using topological features in ASD detection (12, 42).

Different machine learning algorithms were tested on the same dataset to obtain the one with the best performance and then compared with other methods (42), and the comparison showed that once an ML model is trained, it can identify the task condition of new trials. Also, the machine learning classifier can have better sensitivity, as even a single feature that is systematically different between the two task conditions is enough to provide good classification performance, showing in study that the ML classifier identified angry–happy contrasts in more brain structures (19 in each hemisphere) than the permutation cluster test.

11.4 DISCUSSION

11.4.1 ADVANTAGES

Machine learning-assisted EEG analysis offers several advantages over traditional approaches: improved accuracy, higher speed, and less time consumption in manual analysis. It is a rapidly growing field that has the potential to significantly improve our understanding of the brain and enable new forms of diagnosis and treatment for neurological and psychiatric disorders. The integration of preprocessing, feature extraction, and classification algorithms automates the workflow and largely reduces the inaccuracy due to visual observation and manual operation. In particular, the combination of different preprocessing algorithms may provide high-quality EEG data. Different machine learning algorithms can be tailored to suit different cognitive tasks and cohorts. The high efficiency of machine learning algorithms enables online and real-time EEG analysis and empowers new applications including BCI and diagnosis of neurological and psychiatric disorders.

11.4.2 LIMITATIONS AND FUTURE DIRECTIONS

Some limitations of the state-of-the-art deserve more attention. First, as aforementioned, the existing datasets are small; there is a scarcity of large public EEG datasets on cognitive task recognition, which may be partly due to the lack of standardization in experimental paradigms and in EEG data collection and preprocessing. The limited sample size will inevitably affect the applicability of the current research works on different cohorts. Second, regarding different machine learning algorithms, the focus of current studies is comparison, not integration. Adaptive preprocessing and classification based on multiple algorithms still need further investigation. It is also observed that many studies were conducted in the laboratory and that a prototype demonstration in operational environment is still an unmet goal in many cases.

Accordingly, improvements can be made in future studies. First, standardized experimental paradigms and protocols can be developed to collect larger datasets and ease the integration of smaller ones. Larger numbers of subjects in wider age ranges may increase the reliability of the findings (8). Second, by integrating different algorithms with adaptive selection and ensemble techniques (e.g., random subspace algorithm), new algorithms may achieve higher efficiency and accuracy (43). In addition, the proposed solutions need be tested in real-world settings. Finally, the application scenarios can be extended. On the one hand, tackling EEG's limited ability to reliably localize sources across subjects may be helpful for generalizing models across subjects (44). On the other hand, effective methods or inferences that is universally applicable to different types of cognitive tasks can also be goals of future research.

REFERENCES

1. Babu AR, Rajavenkatanarayanan A, Brady JR, Makedon F, Assoc Comp M, editors. Multimodal Approach for Cognitive Task Performance Prediction from Body Postures, Facial Expressions and EEG Signal. Workshop on Modeling Cognitive Processes from Multimodal Data (MCPMD); 2018. 2016 October 16; Boulder, CO2016.

2. Wei J, Salvendy G. The Cognitive Task Analysis Methods for Job and Task Design: Review and Reappraisal. Behaviour & Information Technology. 2004;23(4):273–299.
3. Sharma PK, Vaish A. Individual Identification based on Neuro-Signal Using Motor Movement and Imaginary Cognitive Process. Optik. 2016;127(4):2143–2148.
4. Ahumada-Mendez F, Lucero B, Avenanti A, Saracini C, Munoz-Quezada MT, Cortes-Rivera C, et al. Affective Modulation of Cognitive Control: A Systematic Review of EEG Studies. Physiology & Behavior. 2022;249:113743.
5. Zhao J, Li K, Xi X. et al. Analysis of Complex Cognitive Task and Pattern Recognition using Distributed Patterns of EEG Signals with Cognitive Functions. Neural Computing & Applications. 2020. https://doi.org/10.1007/s00521-020-05439-9
6. Gevins A, Chan CS, Jiang A, Sam-Vargas L. Neurophysiological Pharmacodynamic Measures of Groups and Individuals Extended from Simple Cognitive Tasks to More "Lifelike" Activities. Clinical Neurophysiology. 2013;124(5):870–880.
7. Wan W, Cui X, Gao Z, Gu Z. Frontal EEG-Based Multi-Level Attention States Recognition Using Dynamical Complexity and Extreme Gradient Boosting. Frontiers in Human Neuroscience. 2021;15.
8. Apicella A, Arpaia P, Frosolone M, Improta G, Moccaldi N, Pollastro A. EEG-based Measurement System for Monitoring Student Engagement in Learning 4.0. Scientific Reports. 2022;12(1).
9. Tamilarasi K, Jawahar A, Senthilkumar G, Shanker NR. Diagnosis of Delusion and Hallucination from Schizophrenia Patient Using RADWT. Journal of Medical Systems. 2019;43(7).
10. Olcay BO, Ozgoren M, Karacali B. On the Characterization of Cognitive Tasks Using Activity-Specific Short-Lived Synchronization between Electroencephalography Channels. Neural Networks. 2021;143:452–474.
11. Xu F, Zheng W, Shan D, Yuan Q, Zhou W. Decoding Spectro-Temporal Representation for Motor Imagery Recognition Using ECoG-based Brain–computer Interfaces. Journal of Integrative Neuroscience. 2020;19(2):259–272.
12. Majumder S, Apicella F, Muratori F, Das K, IEEE, editors. Detecting Autism Spectrum Disorder Using Topological Data Analysis. IEEE International Conference on Acoustics, Speech, and Signal Processing; 2020. May 4–8; Barcelona, SPAIN2020.
13. Hassan M, Dufor O, Merlet I, Berrou C, Wendling F. EEG Source Connectivity Analysis: From Dense Array Recordings to Brain Networks. PLoS ONE. 2014;9(8).
14. Amin HU, Mumtaz W, Subhani AR, Saad MNM, Malik AS. Classification of EEG Signals Based on Pattern Recognition Approach. Frontiers in Computational Neuroscience. 2017;11.
15 .Curran E, Sykacek P, Stokes M, Roberts SJ, Penny W, Johnsrude I, et al. Cognitive Tasks for Driving a Brain–computer Interfacing System: A Pilot Study. IEEE Transactions on Neural Systems and Rehabilitation Engineering. 2004;12(1):48–54.
16. Rizkallah J, Benquet P, Kabbara A, Dufor O, Wendling F, Hassan M. Dynamic Reshaping of Functional Brain Networks During Visual Object Recognition. Journal of Neural Engineering. 2018;15(5).
17 .Tanveer M, Gupta A, Kumar D, Priyadarshini S, Chakraborti A, Mallipeddi R, et al., editors. Cognitive Task Classification Using Fuzzy based Empirical Wavelet Transform. IEEE International Conference on Systems, Man, and Cybernetics (SMC); 2018. October 7–10; IEEE Syst Man & Cybernet Soc, Miyazaki, JAPAN2018.
18. Zhang J, Yin Z, Wang R. Design of an Adaptive Human-Machine System Based on Dynamical Pattern Recognition of Cognitive Task-Load. Frontiers in Neuroscience. 2017;11.
19. Zhang J, Yin Z, Wang R. Pattern Classification of Instantaneous Cognitive Task-load Through GMM Clustering, Laplacian Eigenmap, and Ensemble SVMs. IEEE-ACM Transactions on Computational Biology and Bioinformatics. 2017;14(4):947–965.

20. Tariq M, Trivailo PM, Simic M. Classification of Left and Right Foot Kinaesthetic Motor Imagery Using Common Spatial Pattern. Biomedical Physics & Engineering Express. 2020;6(1).

21. Portoles O, Blesa M, van Vugt M, Cao M, Borst J. Thalamic Bursts Modulate Cortical Synchrony Locally to Switch between States of Global Functional Connectivity in a Cognitive Task. PLoS Computational Biology. 2022;18(3).

22. Rus ID, Marc P, Dinsoreanu M, Potolea R, Muresan RC, editors. Classification of EEG signals in an Object Recognition task. 13th IEEE International Conference on Intelligent Computer Communication and Processing (ICCP); 2017. September 7–9; Cluj Napoca, ROMANIA2017.

23. Kumari P, Vaish A. Information-Theoretic Measures on Intrinsic Mode Function for the Individual Identification Using EEG Sensors. IEEE Sensors Journal. 2015;15(9):4950–4960.

24. Petrutiu VM, Palcu LD, Lemnaur C, Dinsoreanu M, Potolea R, Mursesan R, et al., editors. Enhancing the Classification of EEG Signals Using Wasserstein Generative Adversarial Networks. IEEE 16th International Conference on Intelligent Computer Communication and Processing (ICCP); 2020. September 3–5; Electr Network2020.

25. Doborjeh ZG, Doborjeh M, Kasabov N, IEEE, editors. EEG Pattern Recognition using Brain-Inspired Spiking Neural Networks for Modelling Human Decision Processes. International Joint Conference on Neural Networks (IJCNN); 2018. July 8–13; Rio de Janeiro, BRAZIL2018.

26. Wang N, Zhang L, Liu G. EEG-based Research on Brain Functional Networks in Cognition. Bio-Medical Materials and Engineering. 2015;26:S1107–S1114.

27. Hosseini MP, Hosseini A, Ahi K. A Review on Machine Learning for EEG Signal Processing in Bioengineering. IEEE Reviews in Biomedical Engineering. 2021;14:204–218.

28. Doborjeh MG, Kasabov N, Doborjeh ZG. Evolving, Dynamic Clustering of Spatio/Spectro-Temporal Data in 3D Spiking Neural Network Models and a Case Study on EEG Data. Evolving Systems. 2017;9(3):195–211.

29. Cabrera AF, Dremstrup K. Auditory and Spatial Navigation Imagery in Brain–computer Interface Using Optimized Wavelets. Journal of Neuroscience Methods. 2008;174(1):135–146.

30. Satyender, Dhull SK, Singh KK. EEG Artifact Removal Using Canonical Correlation Analysis and EMD-DFA based Hybrid Denoising Approach. Procedia Computer Science. 2023;218:2081–2090.

31. Jiang X, Bian GB, Tian Z. Removal of Artifacts from EEG Signals: A Review. Sensors (Basel). 2019;19(5).

32. Albera L, Kachenoura A, Comon P, Karfoul A, Wendling F, Senhadji L, et al. ICA-based EEG Denoising: A Comparative Analysis of Fifteen Methods. Bulletin of the Polish Academy of Sciences-Technical Sciences. 2012;60(3):407–418.

33. Janani AS, Grummett TS, Lewis TW, Fitzgibbon SP, Whitham EM, DelosAngeles D, et al. Improved Artefact Removal from EEG Using Canonical Correlation Analysis and Spectral Slope. Journal of Neuroscience Methods. 2018;298:1–15.

34. Islam MK, Rastegarnia A, Yang Z. Methods for Artifact Detection and Removal from Scalp EEG: A Review. Clinical Neurophysiology. 2016;46(4–5):287–305.

35. Mumtaz W, Rasheed S, Irfan A. Review of Challenges Associated with the EEG Artifact Removal Methods. Biomedical Signal Processing and Control. 2021;68.

36. .Islam MK, Rastegarnia A, Yang Z. Methods for Artifact Detection and Removal from Scalp EEG: A Review. Neurophysiologie Clinique/Clinical Neurophysiology. 2016;46(4):287–305.

37. Uriguen JA, Garcia-Zapirain B. EEG Artifact Removal-State-of-the-Art and Guidelines. Journal of Neural Engineering. 2015;12(3):031001.

38. Pooja U, Pahuja SK, Veer K. Recent Approaches on Classification and Feature Extraction of EEG Signal: A Review. Robotica. 2021;40(1):77–101.
39. Pooja U, Pahuja SK, Veer K. Recent Approaches on Classification and Feature Extraction of EEG Signal: A Review. Robotica. 2022;40(1):77–101.
40. Olcay BO, Karacali B. Evaluation of Synchronization Measures for Capturing the Lagged Synchronization between EEG Channels: A Cognitive Task Recognition Approach. Computers in Biology and Medicine. 2019;114.
41. Aggarwal S, Chugh N. Review of Machine Learning Techniques for EEG Based Brain Computer Interface. Archives of Computational Methods in Engineering. 2022;29(5):3001–3020.
42. Donos C, Blidarescu B, Pistol C, Oane I, Mindruta I, Barborica A. A comparison of uni- and multi-variate methods for identifying brain networks activated by cognitive tasks using intracranial EEG. Frontiers in Neuroscience. 2022;16.
43. Yang S, Yin Z, Wang Y, Zhang W, Wang Y, Zhang J. Assessing cognitive mental workload via EEG signals and an ensemble deep learning classifier based on denoising autoencoders. Computers in Biology and Medicine. 2019;109:159–170.
44. Courellis H, Mullen T, Poizner H, Cauwenberghs G, Iversen JR. EEG-Based Quantification of Cortical Current Density and Dynamic Causal Connectivity Generalized across Subjects Performing BCI-Monitored Cognitive Tasks. Frontiers in Neuroscience. 2017;11.
45. Keirn ZA, Aunon JI. A New Mode of Communication between Man and His Surroundings. IEEE Transactions on Biomedical Engineering. 1990;37(12):1209–1214. https://doi.org/10.1109/10.64464

12 Detection of Stress Levels During the Stroop Color-Word Test Using Multivariate Projection-Based MUSIC Domain EWT of Multichannel EEG Signal and Machine Learning

Shaswati Dash, Rajesh Kumar Tripathy,
Satrujit Mishra, and Ram Bilas Pachori

12.1 INTRODUCTION

The Stroop Color-Word Test (SCWT) is used in neuroscience applications to evaluate the cognitive inference ability during processing multiple stimuli [1]. This test has been used to induce stress in the subject and is also considered for the diagnosis of attention deficit hyperactive disorder (ADHD) [2, 3]. In SCWT, different printed color patches are shown to the subjects, and the subjects need to identify the colors from those color patches [1]. Typically, two conditions, congruent and incongruent, are considered for SCWT.

The subjects must read the color names and identify the color patches in the congruent condition. Similarly, during the incongruent condition of the SCWT, the subjects must name the color of the ink. The SCWT has been used to quantify various cognitive functions such as stress level, attention, cognitive flexibility, working memory, and processing speed [1]. The study of the induced stress levels during SCWT is interesting in cognitive neuroscience applications. Artificial intelligence (AI)-based algorithms have been used for stress, and emotion recognition using speech and physiological signals [4]. The multichannel EEG signal is widely used for applications such as emotion recognition [5], cognitive task recognition [6], sleep monitoring [7], brain–computer interface (BCI) [8], and diagnosis of various neurological ailments

DOI: 10.1201/9781003479970-12

[**9, 10**]. The automated recognition of stress levels from multichannel EEG signal using AI-based methods is vital to measure the cognitive workload during SCWT.

In literature, various methods have been proposed to detect stress during SCWT using EEG signals [4]. Hamid et al. [11] evaluated EEG signal's power spectral density (PSD) features to discriminate the stress level from the normal state. Similarly, in another study, authors extracted discrete wavelet transform (DWT) based features from EEG signals and used the support vector machine (SVM) model for stress recognition during SCWT [12]. Hou et al. [13] considered statistical features, fractal dimension, and PSD-based features of multichannel EEG signals for stress recognition using an SVM classifier. Similarly, Alonso et al. [14] used the functional connectivity and spectral power-based features of multichannel EEG signals for stress assessment during SCWT.

Similarly, Lim and Chia [15] evaluated the discrete cosine transform coefficients from the single channel EEG signals recorded during SCWT. They used k-nearest neighbor (KNN) and neural network-based classifiers for cognitive stress level recognition. Ajay et al. [16] used DWT to evaluate rhythms from the EEG signal. They extracted PSD-based features from the theta, alpha, and beta rhythms of the EEG signal and used the SVM classifier to classify stress and non-stress during SCWT.

Prerna et al. [17] extracted statistical and Hjorth parameters from the multichannel EEG signal. They used the SVM classifier for two-level and three-level mental stress recognition. Guo et al. [18] used the SWCT to induce stress and extracted PSD features from each rhythm of the EEG signal. They used an SVM classifier for the automated detection of three-level stress. Furthermore, Dorota et al. [19] performed stress-level classification using multilayer perceptron and PSD features of EEG signals during virtual reality. A few methods have also been reported for the automated detection of cognitive workload using EEG signals [20–22]. In [20], the authors computed entropy features in the stationary wavelet transform domain of EEG signal and used the SVM classifier for cognitive workload detection.

Similarly, in [21], and [22], the authors used circulant singular spectrum analysis followed by optimization-based feature selection and SVM classifier for cognitive task classification using EEG signals. The authors of these studies used EEG signals from a few subjects to evaluate their algorithms' performance for identifying stress levels using SCWT. Furthermore, they considered subject-dependent validation schemes to evaluate the classifiers' performance for stress-level recognition. The authors considered univariate signal processing-based methods to evaluate features from small numbers of EEG signal channels for stress-level identification during the SCWT. Therefore, methods based on multivariate signal analysis and the classification models formulated using subject-independent cross-validation (CV) can be developed to detect stress levels using multichannel EEG signal during SCWT automatically.

Multivariate projection (MP)-based empirical wavelet transform (EWT) has been recently proposed to decompose the multichannel multi-component signal into components or modes along all channels [7]. This method has been used to classify different types of sleep stages using features obtained from the multichannel EEG signals [7]. In the MP-EWT method, the discrete Fourier transform (DFT) is used to evaluate the spectrum of the projected EEG signal to detect boundary points and design the EWT filter-bank [7, 23]. The projected EEG signal is calculated by projecting the multichannel EEG data into a unit vector defined using the direction cosine. Once the filter bank is designed using the DFT projected signal, it is used to evaluate the modes of each channel EEG signal [7].

The multiple signal classification (MUSIC) algorithm is an eigenvector-based approach for the spectral estimation of the signals, and it has a better spectral resolution than DFT [24, 25]. The MUSIC-based spectrum of the projected EEG signal can be computed and it can be utilized to detect boundary points using the detection of local maxima. The EWT filter bank can be implemented based on the boundary points of the MUSIC-based spectrum of the projected EEG signals. Similarly, the decision tree (DT) and extended versions such as extreme random tree (ERT), random forest (RF), cross-gradient boosting (XGBoost), and light gradient boosting (LGBM) models have been used for different pattern-classification applications [26–28]. The MP-EWT-based filter bank has yet to be explored to recognize stress levels using multichannel EEG signals during the SCWT. The novelty of this work is the development of the MP-based MUSIC domain EWT (MP-MUSIC-EWT) filter bank for stress-level recognition applications using multichannel EEG signals. The salient contributions of this paper are as follows:

- The MP-MUSIC-EWT filter bank is introduced to decompose the multi-channel EEG signal into modes.
- The attention entropy (AE) and Hjorth parameter-based features from each mode of multichannel EEG signal are evaluated.
- The multilinear singular value decomposition (MLSVD) based tensor factorization is employed to evaluate the features from the third-order tensor (samples×channels×modes) of multichannel EEG signals.
- The DT, RF, and gradient boosting-based models are used for stress-level recognition during SCWT using subject-independent and subject-dependent cases.

The remaining sections of this paper are structured as follows. In Section 2, we have written the details about the SCWT stimuli-based multichannel EEG signal database. The proposed method is described in Section 3. The results and the discussion are written in Section 4. Finally, the conclusions are drawn in Section 5.

12.2 MULTICHANNEL EEG DATABASE

This study uses the multichannel EEG signals from a publicly available database (SAM-40) [29] to evaluate the proposed filter-bank-based approach (Figure 12.1) for cognitive stress-level recognition. In this database, the subject's multichannel EEG signal data is recorded while performing the mirror image recognition, arithmetic calculation, and SCWT-based cognitive tasks [29]. The signals are collected from 40 subjects (26 males and 14 females with an average age of 21.5 years), and the sampling frequency of each EEG signal is 128 Hz.

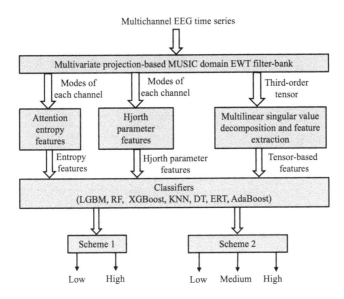

FIGURE 12.1 Flow chart of the proposed MP-MUSIC-EWT-based approach for SCWT-based recognition of cognitive tasks.

Thirty-two channels are used for the recording of the multichannel EEG signals. The dynamic range and the number of bits to store the multichannel EEG signals are given in the database as $\pm 4.1 mV$ and 14 bits, respectively [29]. The duration of multichannel EEG recording for each subject is 25 s for each cognitive task. Three trials are associated with the multichannel EEG recordings of each subject for different cognitive tasks. Both raw and filtered multichannel EEG signals are given in the database [29].

The filtered EEG signal for each channel has been evaluated using the Savitzky–Golay filter followed by the wavelet thresholding [30]. In this work, using the proposed method, we used the 25 s duration of multichannel EEG signals from all subjects for stress-level recognition. The database gives stress-level scores in the range of 1 to 10 for each multichannel EEG signal trial. In this study, we considered a stress-level score of less than 5 to reflect low stress and a score of 5 or higher to indicate a high stress level for the two-class recognition task during the SCWT. Similarly, for the three-class stress-level recognition task during SCWT, we divided the score range into three classes: low (1–3) versus medium (4–6) versus high (7–10) stress.

12.3 PROPOSED METHOD

The flow chart for the stress-level recognition during the SCWT using the proposed approach is depicted in Figure 12.1. It consists of three stages: the MP-MUSIC-EWT for decomposing multichannel EEG signal into modes, the extraction of entropy and

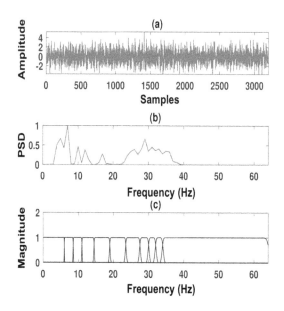

FIGURE 12.2 (a) Plot for the combined EEG (CE) signal. (b) MUSIC-based PSD plot of CE signal. (c) MP-MUSIC-EWT filter-bank designed using MUSIC-based PSD of CE signals.

tensor domain features, and using different machine learning models for stress-level recognition. The descriptions of each stage are given in Figure 12.2.

12.3.1 MULTIVARIATE PROJECTION-BASED MUSIC DOMAIN EWT

MP-EWT is a multivariate signal processing method of decomposing multichannel non-stationary discrete-valued signal data into modes [7]. It mainly evaluates a combined or composite signal from the multichannel signal. The EWT filter bank is designed based on the detected peaks of the Fourier spectrum of the combined signal [7]. The modes of all channel signals are evaluated using a filter bank designed from the frequency domain representation of the combined signal [7].

MP-EWT uses the Fourier spectrum to design the EWT filter bank. EEG signals are nonstationary [31]; therefore, model-based spectral estimation can be used to evaluate the PSD of this signal [24]. The procedure for designing the proposed MP-MUSIC-EWT filter bank for decomposing multichannel EEG signal is given as follows. The multichannel EEG signal is denoted as $X \in \mathfrak{R}^{M \times N}$ with $X = \left[x^m \left(n \right) \right]_{m=1,n=1}^{M,N}$ where M and N are the numbers of channels and samples of multichannel EEG data. The composite EEG signal is calculated by projecting the multichannel EEG data into the M dimensional unit vector on $(M\text{-}1)$ dimensional unit sphere [32]. The direction cosines of the M dimensional unit vector are given by $\dfrac{1}{\sqrt{M}}$. Thus, the unit vector can be written as follows [32]:

$$\tilde{e} = C_1 \tilde{e}_1 + C_2 \tilde{e}_2 + \cdots + C_M \tilde{e}_M \tag{1}$$

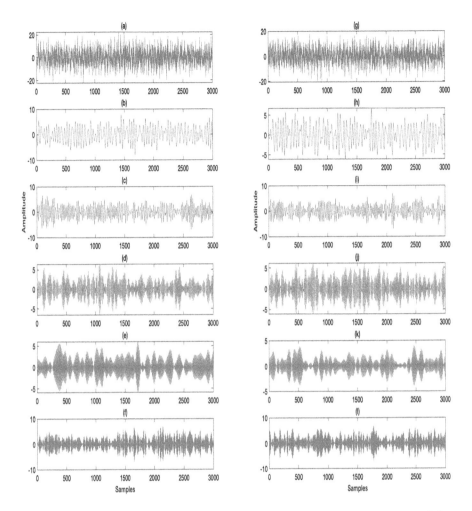

FIGURE 12.3   (a)Fp1 channel EEG signal for SCWT-based low-stress level class. (b)–(f) first five modes of Fp1 channel EEG signal for SCWT-based low-stress-level class. (g) Fp1 channel EEG signal for SCWT-based high-stress-level class. (h)–(l) first five modes of Fp1 channel EEG signal for SCWT-based high-stress-level class.

where $\left[C_1, C_2, \cdots\cdots, C_m \right]$ are the direction cosines, and for this work we used $C_m = \dfrac{1}{\sqrt{M}}$ where $m = 1, 2, \ldots, M$, and $M = 32$ is the total number of EEG channels. The combined EEG signal, $CE(n)$, is evaluated as follows [7]:

$$CE(n) = \frac{1}{\sqrt{M}} \sum_{m=1}^{M} x^m(n)\, \tilde{e}_m \qquad (2)$$

where $x^m(n)$ is denoted as the EEG signal for m^{th} channel. The MUSIC algorithm-based pseudo spectrum of the CE signal is evaluated as follows [24]:

$$P_{MUSIC}^{CE}\left(e^{\frac{j2\pi f}{N}}\right) = \frac{1}{\sum_{l=p+1}^{\theta}\left|V_l S\left(e^{\frac{j2\pi f}{N}}\right)\right|^2} \tag{3}$$

where V_l corresponds the lth eigenvector of the correlation matrix, $S\left(e^{\frac{j2\pi f}{N}}\right)$ is the complex exponential for f^{th} frequency component.

Similarly, θ represents the number of sinusoids; in this work, we set θ at 60 in. The MUSIC algorithm can produce a high-resolution spectral representation of the signal with better localization of peaks in the presence of noise [25]. In this study, we detected peaks in the PSD of the CE signal evaluated using the MUSIC algorithm. Once the peaks are detected, the locations of the peaks are evaluated. The average of the locations of the two adjacent peaks is used to evaluate the frequency points [24]. These frequency points (F_i) are used to evaluate the boundary points (B_i) for the design of the EWT filter bank. The relationship between i^{th} boundary point (B_i) and i^{th} frequency point is given as follows [7, 25]:

$$B_i = \frac{2\pi F_i}{F_s} \tag{4}$$

After evaluating the boundary points, the segments in the spectrum of the combined signals are evaluated as follows:

$$Se_i = [B_{i-1}, B_i] \tag{5}$$

where $B_0 = 0$, and $B_T = \frac{F_s}{2}$ are lower and upper boundary points and $i = 1,2,\ldots, T$. The first segment in the spectrum of the CE signal is $Se_1 = [0, B_1]$ and subsequently, other segments are denoted by $Se_i = [B_{i-1}, B_i]$ with $i = 2,3, \ldots, T$. The empirical scaling function is designed using the first segment $[0, B_1]$, and it is given as follows [23]:

$$\varnothing_1(k) = \begin{cases} 1 & ; if\,|f| \le B_1 - \gamma_1 \\ \cos\left(\frac{\pi}{2}h\left(\frac{1}{2\gamma_1}\left(|f| - B_1 + \gamma_1\right)\right)\right) & ; if\, B_1 - \gamma_1 \le |f| \le B_1 + \gamma_1 \\ 0 & ; otherwise \end{cases} \tag{6}$$

Similarly, the empirical wavelet function for i^{th} segment is given as follows [7, 23]:

$$\varphi_i(k) = \begin{cases} 1 & ; if\ B_i + \gamma_i \leq |f| \leq B_i - \gamma_i \\ \cos\left(\dfrac{\pi}{2} h\left(\dfrac{1}{2\gamma_i + 1}(|f| - B_{i+1} + \gamma_{i+1})\right)\right) & ; if\ B_{i+1} - \gamma_{i+1} \leq |f| \leq B_{i+1} + \gamma_{i+1} \\ \sin\left(\dfrac{\pi}{2} h\left(\dfrac{1}{2\gamma_i + 1}(|f| - B_i + \gamma_i)\right)\right) & ; if\ B_i - \gamma_i \leq |f| \leq B_i + \gamma_i \\ 0 & ; otherwise \end{cases} \tag{7}$$

where the factor $h(r) = 35r^4 - 84r^5 + 70r^6 - 20r^7$. Similarly, the factor $2\gamma_i$ is the transition phase width of ith boundary point, and $\gamma_i = \alpha B_i$, where $\alpha < min_i\left(\dfrac{B_{i-1} - B_i}{B_{i+1} - B_i}\right)$ with $0 < \alpha < 1$ [23]. After the design of the filter bank from the boundary points from the MUSIC-based power spectrum of the CE signal, the same filter bank is utilized to evaluate the modes of each channel EEG signal. The CE signal is depicted in Figure 12.2 (a). Similarly, the MUSIC PSD of the CE signal is shown in Figure 12.2 (b). The filter-bank designed using the MP-MUSIC-EWT-based approach is depicted in Figure 12.2 (c). The first mode is evaluated as follows [7]:

$$mode_1^m(n) = R\left[\frac{1}{N}\sum_{k=0}^{N-1}(x^m(k)\varnothing_1(k))e^{\frac{j2\pi nk}{N}}\right] \tag{8}$$

Similarly, the modes from $i = 2, 3, \ldots, T$ are evaluated as follows [7]:

$$mode_i^m(n) = R\left[\frac{1}{N}\sum_{k=0}^{N-1}(x^m(k)\varphi_i(k))e^{\frac{j2\pi nk}{N}}\right] \tag{9}$$

where $x^m(k)$ is denoted the spectrum or DFT of mth channel EEG signal $x^m(n)$. The EEG signals of Fp1 channels for low-stress level and high-stress level classes during SCWT are shown in Figure 12.3 (a) and Figure 12.3 (b), respectively. Similarly, the first five modes evaluated using MP-MUSIC-EWT filter-bank of EEG signals for both low-stress level and high-stress level classes are depicted in Figure 12.3 (b)-(f), and Figure 12.3 (h)-(l), respectively. The amplitude values for each sample of modes are different for low-stress and high-stress levels. Hence, the features extracted from these modes of different channel EEG signals can be utilized for the automated recognition of stress levels during SCWT.

Algorithm 1: Algorithm of MP-MUSIC-EWT filter bank to evaluate modes of multichannel EEG signal

	Input: Multichannel EEG signal, $X \in \mathfrak{R}^{M \times N}$, where $m \to channels$ and $n \to samples$.
	Output: A third-order tensor containing modes of all channels, $Z \in \mathfrak{R}^{m \times n \times p}$, where $p \to modes$.
Step 1	Compute the combined signal from multichannel signal using equation (2).
Step 2	Evaluate the PSD of the combined signal using the MUSIC-based power spectral estimation method in equation (3).
Step 3	Detect the peaks in the PSD representation of the combined signal and find out the locations of all detected peaks.
Step 4	The average of the two adjacent peak locations in the PSD representation of the combined EEG signal is used as the boundary points to design an EWT filter bank using equation (6) and equation (7).
Step 5	Evaluate the modes of all channel EEG signals using the designed EWT filter bank from the PSD of combined EEG using equation (8) and equation (9).

The third-order tensor of the multichannel EEG signal obtained after the MP-MUSIC-EWT phase is given by $Z \in \mathfrak{R}^{m \times n \times p}$. This work employs the MLSVD to decompose the third-order tensor into a core tensor and factor matrices [33]. The MLSVD of the tensor is given as follows [33, 34]:

$$Z = \Im \times_1 U \times_2 V \times_3 W \tag{10}$$

where $\Im \times_n \{U, V, W\}$ is interpreted as the n-mode product between tensor $(\Im)$ and factor matrices. $\Im = \left\{ \Im_{k_1, k_2, k_3} \right\}$ is the core tensor containing n-mode singular values. In this study, we used all elements of the core tensor $(\Im)$ along three directions (modes, channels, and samples) as features. Similarly, we evaluated the features extracted from the orthogonal matrices as follows:

$$feat_u = \max \left(\max \left(U \right) \right) \tag{11}$$

$$feat_v = \max \left(\max \left(V \right) \right) \tag{12}$$

$$feat_w = \max \left(\max \left(W \right) \right) \tag{13}$$

In this study, the AE features are evaluated from each mode of multichannel EEG signal. The AE considers the local maxima and minima information to capture non-linearity and randomness of the signal data [35]. The modes evaluated in the MP-MUSIC-EWT stage are nonstationary and capture the local information of multichannel EEG signals. The AE features computed from these modes can capture the complexity or randomness of local components or modes of multichannel EEG signals. The ith mode of mth channel EEG signal is given by $mode_i^m (n)$. The AE is evaluated using four steps [35]. First, the local maxima and local minima points are evaluated from the ith mode of mth channel EEG signal. These points are evaluated using the criteria as follows [35]:

1. If $\{mode_i^m(n-1) < mode_i^m(n)\}$ and $\{mode_i^m(n) < mode_i^m(n+1)\}$, then $mode_i^m(n)$ is a local maxima.
2. If $\{mode_i^m(n) < mode_i^m(n-1)\}$ and $\{mode_i^m(n) < mode_i^m(n+1)\}$, then $mode_i^m(n)$ is a local minima.

The signal for i^{th} mode of mth channel EEG can be evaluated by considering the signal of peak points.

Second, the interval between the successive peak points of the ith mode of mth channel is evaluated using the following four ways. Max-max is the interval between the local maxima of each mode of the EEG signal and is denoted as $d1_i^m(k)$ [35]. Min-min is the duration between local minima for each mode of the EEG signal and is given as $d2_i^m(k)$. Likewise, max-min is the interval between local maxima and local minima for each mode is evaluated and denoted by $d3_i^m(k)$. Furthermore, the duration between local minima and local maxima for each mode is computed and it is given as $d4_i^m(k)$. The Shannon entropy of the distributions of all four distance time series for i^{th} mode of mth channel EEG are evaluated and these are denoted as $SE1_i^m$, $SE2_i^m$, $SE3_i^m$, and $SE4_i^m$, respectively. The AE of ith mode of mth channel EEG signal is written as follows [35]:

$$AE_i^{th} = \frac{SE1_i^m + SE2_i^m + SE3_i^m + SE4_i^m}{4} \tag{14}$$

Furthermore, we evaluated the Hjorth parameter [36] from i^{th} mode of mth channel EEG signal. In this work, we computed 320 AE and 320 Hjorth parameter features from all modes of multichannel EEG signals. We evaluated the performance of the classifiers using 365-dimensional tensor domain features, 320-dimensional AE-based features, 320-dimensional Hjorth parameter-based features, and 685-dimensional combined features.

12.3.2 Machine Learning Models

In this work, we used eight machine learning-based classifiers for the automated recognition of stress levels during the SCWT using the MP-MUSIC-EWT domain features of multichannel EEG signals. These classifiers are KNN [37], DT with Gini index, DT with entropy [37], adaptive boosting (AdaBoost) [38], RF [39], ERT [28], XGBoost [40], and LGBM [41]. KNN is a distance-based supervised machine learning algorithm that uses the labels of the K nearest neighbors of training feature vectors based on the minimum distance criteria to predict the outcome of each feature vector in the test feature matrix [37].

Similarly, DT employs splitting criteria, tree depth, and the number of sample leaf parameters to produce an interpretable model that can be used to compute the class labels from the feature vectors from the test feature matrix [37]. RF comprises an ensemble of different DT-based models [39], and the output of each DT model and majority voting criteria are utilized to determine the class label of the feature vector from the testing feature matrix. In ERT, randomized DTs are fitted to the various sub-samples of the feature matrix [28], and the class label for the test instance feature vector is determined using the average of the outputs of all randomized DTs. In contrast to RF, ERT has the benefit of regulating the over-fitting problems [28].

AdaBoost is an ensemble-learning-based method in which the number of weak learners is used to produce a strong classifier to improve the classification accuracy [38]. This study uses decision trees with one level as weak learners for the AdaBoost model. Like the DT-based ensemble model, XGBoost also provides the advantages of regularization, tree pruning, and the optimum gradient boosting method for different supervised learning-based EEG signal processing applications [26, 40]. In contrast to XGBoost, which considers the rise of DT level-wise, LGBM divides the DT leaf-wise and has less computational complexity than XGBoost [41].

The feature matrix evaluated using all three trials' multichannel EEG signals from different subjects is denoted as $F \in \mathfrak{R}^{p \times q}$, where p and q are the total numbers of trails or instances and features of multichannel EEG signals. The classification strategies—low versus high stress and low versus medium versus high—are formulated using the features from multichannel EEG signals using classification models. The subject-independent and subject-dependent validation methods are used for selecting the training and testing instances of the machine learning models. For subject-dependent validation, we considered 80%, and 20% instances or trials of multichannel EEG recordings as training, and testing of the classifiers for two-class and three-class SCWT-based stress level recognition schemes [7].

Similarly, for subject-independent validation case, we used leave-one-out cross-validation where the feature vectors of the multichannel EEG signal trials for one subject are used for the testing of the classifier [6]. The feature vectors of the multichannel EEG signal trails of all other subjects are utilized for the training of the classifier. The metrics such as accuracy, sensitivity, specificity, and F1-score [42] were used to evaluate the classification performance of all eight machine learning-based classifiers for two-class and three-class stress-level recognition tasks during SCWT using multichannel EEG signal features.

The optimal parameters for all eight classifiers are selected using a grid-search-based method with 5-fold cross-validations [26, 43]. For KNN, the number of nearest neighbors grid is selected between 5 to 40 with an increment of 5 [43]. The optimal nearest neighbors obtained after grid-search is 10. Similarly, for DT with entropy and DT with Gini-index models, the grids for maximum depth and the number of leaves are selected from 1 to 6 with an increment of 1. The optimal maximum depth and number of leaves are both 6.

For RF, the grid for the number of trees is selected between 5 to 40 with an increment of 5. The grids for maximum depth and the number of leaves are chosen the same as the DT models. The optimal parameters of the RF model, such as the number of trees, maximum depth, and number of leaves, are obtained as 10, 2, and

1, respectively. Similarly, for the ERT classifier, the grid for the number of trees is selected between 5 to 30 with an increment of 5. The grids for the maximum depth and number of leaves are the same as those with RF.

The optimal parameters of ERT are evaluated as the number of trees as 15, maximum depth as 2, and the number of leaves as 1, respectively. Moreover, for XGBoost, the grids for maximum depth and minimum child weight are selected between 0 to 6 with an increment of 1 [43]. Similarly, the learning rate grid is selected between 0 to 0.6 with an increment of 0.1. The optimal maximum depth, minimum child weight, and learning rate are selected as 6, 1, and 0.3, respectively. For LGBM, the learning rate and maximum depth grids are selected as [0.10, 0.15, 0.18, 0.2], and depth = [1, 2, 3, 4, 5], respectively [43]. The optimal learning rate and maximum depth values are obtained as 0.15 and 5, respectively. Similarly, for AdaBoost, the number of trees grid is between 5 to 50 with an increment of 5. The learning rate grid is chosen as [0.1, 0.5, 1.0, 1.5] [43]. The optimal parameters for AdaBoost are obtained as 10 and 1, respectively.

12.4   RESULTS AND DISCUSSION

The classification results evaluated for two-class and three-class-based stress level recognition schemes using all eight classifiers for subject-dependent and subject-independent validation schemes are presented in this section. In Table 12.1, we show the results for low- versus high stress level recognition tasks using tensor-based, AE-based, combined features and Hjorth parameter-based features evaluated from the MP-MUSIC-EWT domain representation of multichannel EEG signals. RF produced accuracy of 50.43% for a two-class stress-level recognition task using tensor-based features, and accuracy was lower than 50% with the other six classifiers for two-class stress-level detection using the tensor domain features of multichannel EEG signals.

Similarly, higher sensitivity, specificity, and F1-score are also observed for RF than for the other machine learning-based models for two-class-based stress-level recognition using multichannel EEG signals. Similarly, XGBoost produced accuracy, sensitivity, specificity, and F1-score of 54.19%, 51.16%, 56.32%, and 0.53 using MP-MUSIC-EWT domain AE features of multichannel EEG signals. The other classifiers demonstrated worse classification performance. It is also evident from Table 12.1 results that XGBoost achieved accuracy and F1-score of 54.42% and 0.53, respectively, using tensor-based and AE-based features extracted from the MP-MUSIC-EWT domain of multichannel EEG signals. The other classifiers obtained lower accuracy than XGBoost for two-class-based stress-level recognition tasks using combined features of multichannel EEG signals. LGBM obtained the highest accuracy and F1-score of 61.12% and 0.61 using the MP-MUSIC-EWT domain Hjorth parameter-based features compared with other features evaluated from multichannel EEG signals for two-class stress recognition tasks during SCWT.

DT with an entropy-based model obtained accuracy, sensitivity, specificity, and F1-score of 34.46%, 34.46%, 67.34%, and 0.13, respectively, using MP-MUSIC-EWT domain tensor-based features of multichannel EEG signals, and LGBM produced accuracy of 44.89% for a three-class-based stress-level recognition scheme during SCWT. The other classifiers showed worse classification performance than LGBM

TABLE 12.1

Classification Results for Low versus High Stress Level Using Different Feature Combinations with Hold-Out Validation

Features used	Classifiers	Accuracy (%)	Sensitivity (%)	Specificity (%)	F1-score
MP-MUSIC-EWT domain tensor-based features	LGBM	49.35 ± 0.28	47.22 ± 0.34	51.45 ± 0.31	0.49 ± 0.00
	RF	50.43 ± 0.11	48.36 ± 0.10	52.59 ± 0.11	0.50 ± 0.00
	XGBoost	49.15 ± 0.54	47.65 ± 0.28	51.63 ± 0.58	0.48 ± 0.01
	KNN	49.26 ± 0.11	47.66 ± 0.08	51.74 ± 0.14	0.48 ± 0.00
	DT with Gini index	48.28 ± 0.43	47.70 ± 0.11	50.98 ± 0.71	0.42 ± 0.02
	DT with entropy	48.30 ± 0.46	47.71 ± 0.12	50.99 ± 0.71	0.42 ± 0.02
	ERT	49.92 ± 0.16	47.77 ± 0.16	51.99 ± 0.15	0.49 ± 0.00
	AdaBoost	49.94 ± 0.22	47.79 ± 0.21	52.00 ± 0.20	0.49 ± 0.00
MP-MUSIC-EWT domain entropy features	LGBM	54.01 ± 0.26	50.85 ± 0.28	50.85 ± 0.28	0.53 ± 0.00
	RF	50.67 ± 0.20	47.45 ± 0.21	47.45 ± 0.21	0.50 ± 0.00
	XGBoost	54.19 ± 0.12	51.16 ± 0.13	51.16 ± 0.13	0.53 ± 0.00
	KNN	50.95 ± 0.16	48.12 ± 0.13	48.12 ± 0.13	0.50 ± 0.00
	DT with Gini index	53.41 ± 0.16	50.41 ± 0.35	50.41 ± 0.35	0.49 ± 0.00
	DT with entropy	53.26 ± 0.41	50.13 ± 0.78	50.13 ± 0.78	0.49 ± 0.00
	ERT	50.34 ± 0.27	46.97 ± 0.29	46.97 ± 0.29	0.50 ± 0.00
	AdaBoost	50.22 ± 0.17	46.83 ± 0.19	46.83 ± 0.19	0.50 ± 0.00
MP-MUSIC-EWT domain tensor-based and entropy features	LGBM	54.03 ± 0.19	51.89 ± 1.96	51.89 ± 1.96	0.53 ± 0.00
	RF	50.76 ± 0.12	47.54 ± 0.13	47.54 ± 0.13	0.50 ± 0.00
	XGBoost	54.42 ± 0.14	51.45 ± 0.17	51.45 ± 0.17	0.53 ± 0.00
	KNN	51.08 ± 0.19	48.22 ± 0.14	48.22 ± 0.14	0.50 ± 0.00
	DT with Gini index	53.38 ± 0.25	51.44 ± 1.43	51.44 ± 1.43	0.49 ± 0.00
	DT with entropy	53.31 ± 0.38	50.23 ± 0.73	50.23 ± 0.73	0.49 ± 0.00
	ERT	50.69 ± 0.20	47.34 ± 0.21	47.34 ± 0.21	0.50 ± 0.00
	AdaBoost	50.30 ± 0.11	46.93 ± 0.13	46.93 ± 0.13	0.50 ± 0.00
MP-MUSIC-EWT domain Hjorth parameter-based features	LGBM	**61.12 ± 0.16**	58.77 ± 0.25	58.77 ± 0.25	0.61 ± 0.00
	RF	56.45 ± 0.11	53.16 ± 0.11	53.16 ± 0.11	0.56 ± 0.00
	XGBoost	60.56 ± 0.46	59.00 ± 0.63	59.00 ± 0.63	0.60 ± 0.00
	KNN	57.97 ± 0.02	54.08 ± 0.03	54.08 ± 0.03	0.58 ± 0.00
	DT with Gini index	55.35 ± 0.09	55.01 ± 0.93	55.01 ± 0.93	0.50 ± 0.01
	DT with entropy	55.35 ± 0.03	55.01 ± 0.78	55.01 ± 0.78	0.50 ± 0.01
	ERT	56.39 ± 0.24	53.26 ± 0.28	53.26 ± 0.28	0.56 ± 0.00
	AdaBoost	56.41 ± 0.08	53.28 ± 0.08	53.28 ± 0.08	0.56 ± 0.00

with MP-MUSIC-EWT domain AE features. Furthermore, when combined features of multichannel EEG signals are used, XGBoost obtains an accuracy of 37.14% for three-class-based stress level recognition tasks during SCWT.

LGBM achieved the highest accuracy of 48.76% using MP-MUSIC-EWT domain Hjorth parameter-based features of multichannel EEG signals. The MP-MUSIC-EWT domain AE and Hjorth parameter-based features of multichannel EEG signals achieved higher accuracy than tensor domain features automated classification stress levels during SCWT. The AE and Hjorth parameter of each mode of multichannel EEG captures the nonlinearity. Hence, these features have demonstrated better classification performance for stress level recognition during SCWT. In Table 12.2, we

TABLE 12.2

Classification Results for Low versus Medium versus High-Stress Level Scheme Using Different Feature Combinations with Hold-Out Validation

Features used	Classifiers	Accuracy (%)	Sensitivity (%)	Specificity (%)	F1-score
MP-MUSIC-EWT domain tensor-based features	LGBM	33.06 ± 0.55	33.06 ± 0.55	66.59 ± 0.17	0.24 ± 0.00
	RF	32.93 ± 0.24	32.93 ± 0.24	66.49 ± 0.08	0.30 ± 0.00
	XGBoost	33.54 ± 0.27	31.26 ± 5.05	66.71 ± 0.07	0.18 ± 0.00
	KNN	32.68 ± 0.52	32.68 ± 0.05	66.30 ± 0.17	0.30 ± 0.00
	DT with Gini index	34.40 ± 0.23	34.40 ± 0.23	67.28 ± 0.12	0.13 ± 0.00
	DT with entropy	34.46 ± 0.12	34.46 ± 0.12	67.34 ± 0.10	0.13 ± 0.00
	ERT	32.51 ± 0.27	32.51 ± 0.27	66.30 ± 0.09	0.30 ± 0.00
	AdaBoost	33.16 ± 0.43	33.16 ± 0.43	66.47 ± 0.18	0.30 ± 0.00
MP-MUSIC-EWT domain entropy features	LGBM	44.89 ± 8.53	39.92 ± 0.32	68.88 ± 0.09	0.31 ± 0.00
	RF	35.74 ± 0.19	35.74 ± 0.19	67.38 ± 0.03	0.33 ± 0.00
	XGBoost	39.93 ± 0.28	39.93 ± 0.28	68.78 ± 0.17	0.28 ± 0.00
	KNN	36.71 ± 0.39	36.71 ± 0.39	68.07 ± 0.06	0.34 ± 0.00
	DT with Gini index	38.37 ± 0.60	38.37 ± 0.60	68.40 ± 0.11	0.27 ± 0.01
	DT with entropy	38.37 ± 0.69	38.37 ± 0.69	68.30 ± 0.13	0.27 ± 0.01
	ERT	35.38 ± 0.37	35.38 ± 0.37	67.17 ± 0.07	0.32 ± 0.00
	AdaBoost	35.38 ± 0.32	35.38 ± 0.32	67.12 ± 0.10	0.32 ± 0.00
MP-MUSIC-EWT domain tensor-based and entropy features	LGBM	36.05 ± 0.37	36.05 ± 0.00	67.37 ± 0.09	0.25 ± 0.00
	RF	35.60 ± 0.33	35.60 ± 0.00	67.08 ± 0.10	0.30 ± 0.00
	XGBoost	37.14 ± 0.32	37.14 ± 0.13	67.72 ± 0.17	0.21 ± 0.01
	KNN	34.76 ± 0.13	34.76 ± 0.13	67.07 ± 0.01	0.31 ± 0.00
	DT with Gini index	36.24 ± 0.20	36.24 ± 0.20	67.60 ± 0.13	0.19 ± 0.01
	DT with entropy	36.21 ± 0.35	36.21 ± 0.35	67.59 ± 0.15	0.19 ± 0.01
	ERT	34.52 ± 0.42	34.52 ± 0.42	66.88 ± 0.08	0.31 ± 0.00
	AdaBoost	34.97 ± 0.21	34.97 ± 0.21	67.01 ± 0.11	0.30 ± 0.00
MP-MUSIC-EWT domain Hjorth parameter-based features	LGBM	**48.76 ± 0.45**	48.76 ± 0.45	72.06 ± 0.27	0.41 ± 0.00
	RF	43.61 ± 0.54	43.61 ± 0.54	70.55 ± 0.08	0.40 ± 0.00
	XGBoost	45.85 ± 0.13	45.85 ± 0.13	71.69 ± 0.08	0.39 ± 0.01
	KNN	47.04 ± 0.38	47.03 ± 0.38	71.74 ± 0.16	0.42 ± 0.00
	DT with Gini index	38.55 ± 0.08	38.55 ± 0.08	68.80 ± 0.08	0.31 ± 0.01
	DT with entropy	38.25 ± 0.63	38.24 ± 0.63	68.99 ± 0.28	0.30 ± 0.01
	ERT	42.90 ± 0.41	42.89 ± 0.40	70.36 ± 0.11	0.39 ± 0.00
	AdaBoost	43.56 ± 0.35	43.56 ± 0.35	70.55 ± 0.11	0.40 ± 0.00

show the classification results of all classifiers for low versus medium versus high SCWT-based stress-level recognition schemes using different feature sets of multi-channel EEG signals.

The subject-independent CV results evaluated using all classifiers for two-class and three-class-based stress-level recognition schemes using features from multichannel EEG signals are shown in Table 12.3. It is noted that for the two-class

TABLE 12.3

Classification Results for Low versus High SCWT (Two-Class Classification) and Low versus Medium versus High SCWT (Three-Class Classification) Using the MP-MUSIC-EWT Domain Features of EEG Signal with Leave-One-Out Subject Independent Validation

Features used	Classifiers	Accuracy (%) for two-class	Accuracy (%) for three-class
MP-MUSIC-EWT domain tensor-based features	LGBM	49.68 ± 2.23	20.58 ± 8.60
	RF	48.76 ± 3.22	34.73 ± 2.72
	XGBoost	50.44 ± 9.74	10.45 ± 7.04
	KNN	57.45 ± 5.40	35.18 ± 2.11
	DT with Gini index	48.71 ± 26.32	6.93 ± 7.78
	DT with entropy	46.10 ± 26.54	6.65 ± 7.58
	ERT	47.70 ± 3.75	34.06 ± 2.93
	AdaBoost	47.52 ± 3.57	33.92 ± 2.64
MP-MUSIC-EWT domain entropy features	LGBM	50.72 ± 6.20	21.38 ± 10.30
	RF	49.18 ± 2.30	35.30 ± 2.57
	XGBoost	**52.24 ± 7.11**	18.86 ± 14.02
	KNN	53.65 ± 5.01	34.54 ± 3.73
	DT with Gini index	46.62 ± 12.60	16.74 ± 12.02
	DT with entropy	45.74 ± 11.94	16.95 ± 12.11
	ERT	48.10 ± 2.60	34.54 ± 2.70
	AdaBoost	48.02 ± 2.60	34.36 ± 2.73
MP-MUSIC-EWT domain tensor-based and entropy features	LGBM	48.66 ± 6.26	28.28 ± 5.25
	RF	48.04 ± 6.46	31.24 ± 5.50
	XGBoost	48.29 ± 9.78	19.92 ± 7.76
	KNN	54.84 ± 5.99	**38.87 ± 3.70**
	DT with Gini index	56.91 ± 20.74	14.92 ± 10.86
	DT with entropy	**57.36 ± 20.87**	14.43 ± 10.17
	ERT	49.39 ± 7.26	36.61 ± 7.77
	AdaBoost	48.65 ± 6.29	32.25 ± 5.72

stress-level recognition case, KNN achieved accuracy of 54.54% using tensor domain features of multichannel EEG signals. Similarly, KNN also obtained higher accuracy for a three-class-based stress-level recognition scheme using tensor domain features of multichannel EEG signals recorded during SCWT.

The DT classifiers with Gini index and entropy features obtained less than 10% accuracy for the three-class stress-level recognition tasks. RF, ERT, and AdaBoost produced more than 30% accuracy for three-class-based stress level recognition using tensor domain features of multichannel EEG signals. Furthermore, KNN coupled with mode AE features of multichannel EEG signals have achieved the average accuracy values of 53.65% and 34.54%, respectively. XGBoost accuracy was 54.43% using MP-MUSIC-EWT domain Hjorth parameter-based features for two-class stress-level classification tasks during SCWT with subject-independent validation.

Similarly, RF produced accuracy of 38.36% using MP-MUSIC-EWT domain Hjorth parameter-based features for three-class stress-level recognition tasks. When all features (combination of a tensor domain and AE) of multichannel EEG signals were used, the KNN classifier produced average accuracy of 38.87% for the three-class-based stress-level recognition scheme. The accuracy of other classifiers was lower than that with KNN models using all features of multichannel EEG signals to detect stress levels during SCWT with subject-independent CV. Similarly, for the two-class stress-level recognition task, DT with entropy obtained accuracy of 57.36% using MP-MUSIC-EWT domain tensor-based and entropy features.

The classification accuracy of the best-performing classifiers was evaluated for both two-class and three-class-based stress-level recognition tasks using different quantization word lengths for multichannel EEG signals, depicted in Figure 12.4.

XGBoost coupled with all features obtained the highest accuracy for the four-word length-based quantization of multichannel EEG signals for the two-class stress-level classification task. Similarly, for the three-class-based task, LGBM coupled with AE features produced the same overall accuracy for word length as 4, 10, and 12 for quantization of multichannel EEG signals. It is also observed that with less quantization word length of multichannel EEG signal, the classification performance of the proposed approach remained the same for both two-class and three-class-based stress-level recognition tasks during SCWT.

Moreover, we compared the classification accuracy obtained using the proposed method with DWT-based features coupled with SVM, Hjorth parameter-based features coupled with SVM, fractal dimension-based features coupled with SVM, and PSD features coupled with the SVM model for two-class and three-class SCWT-based stress-level recognition tasks in Table 12.4 for the same database multichannel EEG signals. The accuracy of MP-MUSIC-EWT domain features coupled with the gradient boosting model is higher than 60% for two-class stress level classification during the SCWT.

However, DWT, fractal dimension-based, Hjorth parameter-based, and PSD features obtained lower accuracy using the SVM classifier than MP-MUSIC-EWT domain features for the two-class stress level recognition task. Similarly, it is seen that for the three-class stress-level recognition task, the proposed MP-MUSIC-EWT domain entropy features coupled with the LGBM model achieved accuracy of more than 48%, in contrast with other features with SVM.

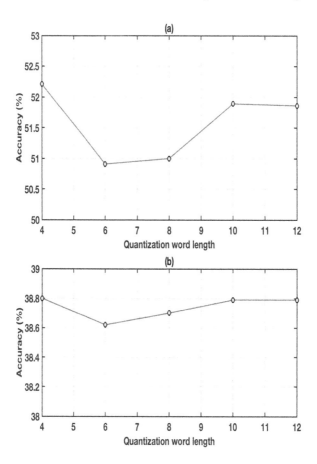

FIGURE 12.4 Quantization word length versus accuracy plot for (a) low versus high stress during the SCWT using combined features of multichannel EEG signals and XGBoost classifier, (b) low versus medium versus high stress levels during the SCWT using entropy features from modes of multichannel EEG signals and LGBM classifier.

TABLE 12.4
Comparison of Performance with Existing Features of EEG Signals for Three-Class and Two-Class Hold-Out Classification Schemes

Features evaluated from multichannel EEG signals	Classifier used	Two-class stress level classification	Three-class stress level classification
		Accuracy (%)	Overall Accuracy (%)
DWT domain features [16]	SVM	47.56	46.17
PSD-based features [11]	SVM	48.35	46.71
Hjorth parameter-based features [17]	SVM	50.71	44.27
Fractal dimension-based features [13]	SVM	54.26	46.35
Proposed work (MP-MUSIC-EWT domain Hjorth parameter-based features)	LGBM	61.12	48.76

We performed analyses of variance (ANOVA) [44] to investigate the significant differences in the accuracy value of the proposed method with existing techniques for two-class and three-class based stress level recognition tasks during SCWT. The p values obtained after ANOVA for two-class and three-class cases are 2.13×10^{-29} and 1.12×10^{-15}, respectively, and the differences in accuracy between the proposed MP-MUSIC-EWT-based approach and existing techniques are significant for the automated recognition of stress levels during SCWT. The advantages of this approach are given as follows:

- The MP-MUSIC-EWT filter bank is introduced in which the EWT filter bank is designed based on the segregation of the MUSIC-based power spectrum of the combined EEG signal.
- The Hjorth parameter and AE-based features from 10 modes and tensor domain features of multichannel EEG signals are evaluated.
- Two-class (low versus high) and three-class (low versus medium versus high)-based stress-level recognition tasks are performed using MP-MUSIC-EWT domain features of multichannel EEG signals.
- The MP-MUSIC-EWT domain Hjorth parameter-based features coupled with LGBM classifier has produced more than 60% accuracy for two-class based stress recognition task than AE and tensor-based features.
- Using multichannel EEG signals, the proposed approach is evaluated for subject-independent and subject-dependent stress level recognition tasks.
- The suggested MP-MUSIC-EWT domain features with the LGBM model perform better than PSD, DWT, Hjorth parameter, and fractal dimension-based features for stress level recognition during the SCWT.

The proposed MP-MUSIC-EWT domain ML-based approach considered the multichannel EEG signals from 40 subjects for stress level recognition during SCWT. In the future, multichannel EEG recordings from more subjects can be recorded and used to test the proposed approach. In this study, we used the multichannel EEG signals from a public database that was developed recently [29]. Our approach is the first to automatically recognize two-class and multiclass based cognitive stress levels using the SAM-40 dataset multichannel EEG signals and has obtained the maximum accuracy value of 57.36% for subject-independent CV cases. The deep learning models such as convolutional neural network [5, 45], time-series transformer [46] and recurrent neural network [6] models can be utilized in the MP-MUSIC-EWT domain modes of multichannel EEG signals to improve the accuracy for stress-level recognition tasks during SCWT. The various multivariate multiscale analysis [34] and time–frequency decomposition techniques [31] can be explored for the automated recognition of stress levels during SCWT using multichannel EEG recordings.

12.5 CONCLUSION

A method for the automated recognition of cognitive stress levels during SCWT using multichannel EEG signals has been proposed in this work. The MP-MUSIC-EWT filter bank has been introduced to evaluate the modes of multichannel EEG signals. The third-order tensor domain features, Hjorth parameter-based features, and AE-based features have been evaluated from the MP-MUSIC-EWT domain modes of multichannel EEG signals. Seven machine learning models (DT, RF, KNN, XGBoost, LGBM, ERT, and AdaBoost) have been used for two-class (low versus high) and three-class (low versus medium versus high) stress-level recognition tasks using multichannel EEG signals recorded during SCWT.

The proposed approach achieved accuracy of 61.12% and 48.76% for two-class and three-class stress-level recognition tasks using MP-MUSIC-EWT domain Hjorth parameter-based features with the LGBM classifier. The XGBoost and LGBM-based machine learning models produced higher accuracy for two- and three-class stress-level recognition schemes. Deep learning and other transform domain machine learning models can be investigated in the future to increase the classification performance of SCWT-based stress level recognition using multichannel EEG signals.

REFERENCES

1. F. Scarpina and S. Tagini, "The stroop color and word test," Frontiers in Psychology, vol. 8, p. 557, 2017.
2. X. Hou, Y. Liu, O. Sourina, and W. Mueller-Wittig, "Cognimeter: EEG-based emotion, mental workload and stress visual monitoring," in 2015 International Conference on Cyberworlds (CW), pp. 153–160, IEEE, 2015.
3. S. Homack and C. A. Riccio, "A meta-analysis of the sensitivity and specificity of the stroop color and word test with children," Archives of clinical Neuropsychology, vol. 19, no. 6, pp. 725–743, 2004.
4. R. Katmah, F. Al-Shargie, U. Tariq, F. Babiloni, F. Al-Mughairbi, and H. Al-Nashash, "A review on mental stress assessment methods using EEG signals," Sensors, vol. 21, no. 15, p. 5043, 2021.
5. D. Maheshwari, S. K. Ghosh, R. Tripathy, M. Sharma, and U. R. Acharya, "Automated accurate emotion recognition system using rhythm-specific deep convolutional neural network technique with multi-channel EEG signals," Computers in Biology and Medicine, vol. 134, p. 104428, 2021.
6. A. Varshney, S. K. Ghosh, S. Padhy, R. K. Tripathy, and U. R. Acharya, "Automated classification of mental arithmetic tasks using recurrent neural network and entropy features obtained from multi-channel EEG signals," Electronics, vol. 10, no. 9, p. 1079, 2021.
7. R. K. Tripathy, S. K. Ghosh, P. Gajbhiye, and U. R. Acharya, "Development of automated sleep stage classification system using multivariate projection-based fixed boundary empirical wavelet transform and entropy features extracted from multichannel EEG signals," Entropy, vol. 22, no. 10, p. 1141, 2020.
8. K. Das and R. B. Pachori, "Electroencephalogram-based motor imagery brain–computer interface using multivariate iterative filtering and spatial filtering," IEEE Transactions on Cognitive and Developmental Systems, vol. 15, no. 3, pp. 1408–1418, September 2023, doi: 10.1109/TCDS.2022.3214081.
9. A. Anuragi, D. S. Sisodia, and R. B. Pachori, "Epileptic-seizure classification using phase-space representation of FBSE-EWT based EEG sub-band signals and ensemble learners," Biomedical Signal Processing and Control, vol. 71, p. 103138, 2022.
10. A. Lenartowicz and S. K. Loo, "Use of EEG to diagnose ADHD," Current Psychiatry Reports, vol. 16, no. 11, pp. 1–11, 2014.
11. F. A. Hamid, M. N. M. Saad, and A. S. Malik, "Characterization stress reactions to stroop color-word test using spectral analysis," Materials Today: Proceedings, vol. 16, pp. 1949–1958, 2019.

12. P. Gaikwad and A. Paithane, "Novel approach for stress recognition using EEG signal by SVM classifier," in 2017 International Conference on Computing Methodologies and Communication (ICCMC), pp. 967–971, IEEE, 2017.
13. X. Hou, Y. Liu, O. Sourina, Y. R. E. Tan, L. Wang, and W. Mueller-Wittig, "EEG based stress monitoring," in 2015 IEEE International Conference on Systems, Man, and Cybernetics, pp. 3110–3115, IEEE, 2015.
14. J. Alonso, S. Romero, M. Ballester, R. Antonijoan, and M. Mañanas, "Stress assessment based on EEG univariate features and functional connectivity measures," Physiological Measurement, vol. 36, no. 7, p. 1351, 2015.
15. C.-K. A. Lim and W. C. Chia, "Analysis of single-electrode EEG rhythms using matlab to elicit correlation with cognitive stress," International Journal of Computer Theory and Engineering, vol. 7, no. 2, p. 149, 2015.
16. A. N. Paithane and M. Alagirisamy, "Electroencephalogram signal analysis using wavelet transform and support vector machine for human stress recognition," Biomedical and Pharmacology Journal, vol. 15, no. 3, pp. 1349–1360, 2022.
17. P. Singh, R. Singla, and A. Kesari, "An EEG based approach for the detection of mental stress level: An application of BCI," in *Recent Innovations in Mechanical Engineering*, pp. 49–57. Springer, 2022.
18. G. Jun and K. G. Smitha, "EEG based stress level identification," in 2016 IEEE International Conference on Systems, Man, and Cybernetics (SMC), pp. 3270–3274, IEEE, 2016.
19. D. Kamińska, K. Smólka, and G. Zwoliński, "Detection of mental stress through EEG signal in virtual reality environment," Electronics, vol. 10, no. 22, p. 2840, 2021.
20. L. D. Sharma, R. K. Saraswat, and R. K. Sunkaria, "Cognitive performance detection using entropy-based features and lead-specific approach," Signal, Image and Video Processing, vol. 15, no. 8, pp. 1821–1828, 2021.
21. J. Yedukondalu and L. D. Sharma, "Cognitive load detection using circulant singular spectrum analysis and binary Harris Hawks optimization based feature selection," Biomedical Signal Processing and Control, vol. 79, p. 104006, 2023.
22. J. Yedukondalu and L. D. Sharma, "Cognitive load detection using binary salp swarm algorithm for feature selection," in 2022 IEEE 6th Conference on Information and Communication Technology (CICT), pp. 1–5, IEEE, 2022.
23. J. Gilles, "Empirical wavelet transform," IEEE Transactions on Signal Processing, vol. 61, no. 16, pp. 3999–4010, 2013.
24. J. P. Amezquita-Sanchez and H. Adeli, "A new music-empirical wavelet transform methodology for time–frequency analysis of noisy nonlinear and non-stationary signals," Digital Signal Processing, vol. 45, pp. 55–68, 2015.
25. K. Agarwal and R. Macháň, "Multiple signal classification algorithm for super-resolution fluorescence microscopy," Nature Communications, vol. 7, no. 1, pp. 1–9, 2016.
26. S. Dash, R. K. Tripathy, D. K. Dash, G. Panda and R. B. Pachori, "Multiscale domain gradient boosting models for the automated recognition of imagined vowels using multichannel EEG signals," IEEE Sensors Letters, vol. 6, no. 11, pp. 1–4, November 2022, Art no. 7004804, doi: 10.1109/LSENS.2022.3218312.
27. R. Tripathy, L. Sharma, and S. Dandapat, "Detection of shockable ventricular arrhythmia using variational mode decomposition," Journal of medical systems, vol. 40, no. 4, pp. 1–13, 2016.
28. P. Geurts, D. Ernst, and L. Wehenkel, "Extremely randomized trees," Machine Learning, vol. 63, no. 1, pp. 3–42, 2006.
29. R. Ghosh, N. Deb, K. Sengupta, A. Phukan, N. Choudhury, S. Kashyap, S. Phadikar, R. Saha, P. Das, N. Sinha, et al., "Sam 40: Dataset of 40 subject EEG recordings to monitor the induced-stress while performing stroop color-word test, arithmetic task, and mirror image recognition task," Data in Brief, vol. 40, p. 107772, 2022.

30. P. Gajbhiye, N. Mingchinda, W. Chen, S. C. Mukhopadhyay, T. Wilaiprasitporn, and R. K. Tripathy, "Wavelet domain optimized Savitzky–Golay filter for the removal of motion artifacts from EEG recordings," IEEE Transactions on Instrumentation and Measurement, vol. 70, pp. 1–11, 2020.

31. Pachori, Ram Bilas. *Time-frequency Analysis Techniques and Their Applications*. CRC Press, 2023.

32. M. R. Thirumalaisamy and P. J. Ansell, "Fast and adaptive empirical mode decomposition for multidimensional, multivariate signals," IEEE Signal Processing Letters, vol. 25, no. 10, pp. 1550–1554, 2018.

33. L. De Lathauwer, B. De Moor, and J. Vandewalle, "A multilinear singular value decomposition," SIAM Journal on Matrix Analysis and Applications, vol. 21, no. 4, pp. 1253–1278, 2000.

34. C. Chauhan, R. K. Tripathy, and M. Agrawal, "Patient specific higher order tensor based approach for the detection and localization of myocardial infarction using 12-lead ECG," Biomedical Signal Processing and Control, vol. 83, p. 104701, 2023.

35. J. Yang, G. I. Choudhary, S. Rahardja and P. Fränti, "Classification of Interbeat Interval Time-Series Using Attention Entropy," IEEE Transactions on Affective Computing, vol. 14, no. 1, pp. 321–330, 1 January–March 2023, doi: 10.1109/TAFFC.2020.3031004.

36. T. Cecchin, R. Ranta, L. Koessler, O. Caspary, H. Vespignani, and L. Maillard, "Seizure lateralization in scalp EEG using Hjorth parameters," Clinical neurophysiology, vol. 121, no. 3, pp. 290–300, 2010.

37. C. M. Bishop and N. M. Nasrabadi, *Pattern Recognition and Machine Learning*, vol. 4. Springer, 2006.

38. R. E. Schapire, "Explaining adaboost," in *Empirical Inference*, pp. 37–52. Springer, 2013.

39. R. M. Mehmood, M. Bilal, S. Vimal, and S.-W. Lee, "EEG-based affective state recognition from human brain signals by using Hjorth-activity," Measurement, vol. 202, p. 111738, 2022.

40. T. Chen and C. Guestrin, "Xgboost: A scalable tree boosting system," in Proceedings of the 22nd ACM SIGKDD International Conference on Knowledge Discovery and Data Mining, pp. 785–794, 2016.

41. G. Ke, Q. Meng, T. Finley, T. Wang, W. Chen, W. Ma, Q. Ye, and T.-Y. Liu, "Lightgbm: A highly efficient gradient boosting decision tree," Advances in Neural Information Processing Systems, vol. 30, 2017.

42. J. Karhade, S. Dash, S. K. Ghosh, D. K. Dash, and R. K. Tripathy, "Time–frequency-domain deep learning framework for the automated detection of heart valve disorders using PCG signals," IEEE Transactions on Instrumentation and Measurement, vol. 71, pp. 1–11, 2022.

43. F. Pedregosa, G. Varoquaux, A. Gramfort, V. Michel, B. Thirion, O. Grisel, M. Blondel, P. Prettenhofer, R. Weiss, V. Dubourg, et al., "Scikit-learn: Machine learning in python," The Journal of Machine Learning Research, vol. 12, pp. 2825–2830, 2011.

44. A. Cuevas, M. Febrero, and R. Fraiman, "An ANOVA test for functional data," Computational Statistics & Data Analysis, vol. 47, no. 1, pp. 111–122, 2004.

45. J. Karhade, S. K. Ghosh, P. Gajbhiye, R. K. Tripathy, and U. R. Acharya, "Multichannel multiscale two-stage convolutional neural network for the detection and localization of myocardial infarction using vectorcardiogram signal," Applied Sciences, vol. 11, no. 17, p. 7920, 2021.

46. A. R. Abbasi, M. R. Mahmoudi, and M. M. Arefi, "Transformer winding faults detection based on time series analysis," IEEE Transactions on Instrumentation and Measurement, vol. 70, pp. 1–10, 2021.

Index